Monographs of the Center for Southeast Asian Studies,
Kyoto University
English-language Series no. 10

PADDY SOILS IN TROPICAL ASIA
THEIR MATERIAL NATURE AND FERTILITY

D. Kummer
1991

Paddy Soils in Tropical Asia

Their Material Nature and Fertility

KEIZABURO KAWAGUCHI

KAZUTAKE KYUMA

*Monographs of the Center
for Southeast Asian Studies
Kyoto University*

THE UNIVERSITY PRESS OF HAWAII
Honolulu

Kawaguchi, Keizaburó, 1914–
   Paddy Soils in Tropical Asia—Their Material Nature and Fertility

   (Monographs of the Center for Southeast Asian Studies, Kyoto
University)
   Bibliography: p.
   Includes index.
   1. Rice—Soils. 2. Soils—South Asia. 3. Soils—Asia, Southeastern.
I. Kyúma, Kazutake, joint author. II. Title. III. Series: Kyōto
Daigaku. Tōnan Ajia Kenkyū Sentā. Monographs of the Center for
Southeast Asian studies, Kyoto University.
S597.R5K38   633′.18   77-21463
ISBN 0-8248-0570-4
ISBN 0-8248-0571-2 pbk.

Filmset in Hong Kong by Asco Trade Typesetting Ltd.
Printed in Hong Kong by South China Photo-Process Printing Co. Ltd.

Contents

Tables

Preface

This book is the outcome of a research project organized by the Center for Southeast Asian Studies of Kyoto University, with the aim of gaining a better understanding of the foundations of rice cultivation, which sustains the way of life of the people of South and Southeast Asia.

Paddy soils or lowland rice soils are the media of food production for many millions of people. In spite of their importance paddy soils have received relatively little attention from soil scientists, because of their localized occurrence and their unusual, at least to western people, management practice of waterlogging. In fact, for rice cultivation water conditions are more important than soil conditions. But this does not mean that soil factors are not relevant to the performance of rice plants.

In Japan paddy soils have been the object of intensive studies for the last 40 years. These studies have clarified many aspects of paddy soils in relation to rice cultivation. However, the acreage of paddy soil in Japan constitutes only a little more than 2% of the world total, and the greater part of the remaining acreage occurring in tropical Asia has been little studied. The most important motivation of the present research was to fill this gap, in particular in relation to the nature and fertility of paddy soils in tropical Asia.

As there is no absolute criterion for judging soil fertility, its evaluation for a large area like tropical Asia has to be carried out on a comparative basis, with standardized methods, and preferably by the same researchers. To our knowledge there have been no previous studies covering a wide area. We believe that our research as presented here is the first attempt of its kind in the history of soil science.

Since 1963, when the first fieldwork was conducted in Thailand, we have made more than 20 surveys, not only in South and Southeast Asia, but also, for comparison, in the rice-growing areas of Australia and in Mediterranean countries. We spent one to three months each time in investigating and collecting samples of paddy soils. The collected samples were analyzed at either the Faculty of Agriculture or the Center for Southeast Asian Studies, Kyoto University.

Some of the earlier results have been reported in two monographs, *Lowland Rice Soils in Thailand* and *Lowland Rice Soils in Malaya*, published by the Center for Southeast Asian Studies in 1969, which are detailed accounts of paddy soils in those two countries. In the present book we have organized our results to enable the reader to locate the soils of one country or region in the wide spectrum of material and fertility characteristics for the paddy soils of tropical Asia. In other words, specific features of the paddy soil of a country or a region are clarified in comparison with soils in other countries or regions within tropical Asia.

In this book no particular attempt has been made to compare paddy soils in tropical Asia with those in the temperate zone. This is because variation within tropical Asia is wider than that between the two climatic regions, and, secondly, the fundamental processes operating in paddy soils, whether physical, chemical, or biological, are essentially the same in the two zones. The greatest difference is often seen in factors influenced by human intervention, for example, in phosphorus content resulting from application of fertilizers.

Although we are aware that there may be room for further improvement, we felt it was high time for us to make some concrete return to those who had encouraged and helped us in so many different ways during the course of our research. And so we present this book as our thanks.

It is impossible to list the names of all the individuals to whom we are deeply indebted. We wish to express our most sincere gratitude to the people who helped us to the utmost during our field surveys in Bangladesh, Burma, Cambodia, India, Indonesia, Malaysia, the Philippines, Sri Lanka, Thailand and Vietnam, and in Australia, Italy, Portugal and Spain.

Our thanks are also due to mentors and colleagues at Kyoto University, in particular those in the Faculty of Agriculture and the Center for Southeast Asian Studies, who encouraged and supported us throughout the research period.

Financial support was obtained from the International Rice Research Institute in the Philippines and the Ministry of Education of Japanese Government, in addition to that received from the Center for Southeast Asian Studies, Kyoto University. We are grateful for the generosity with which these institutions donated their precious funds for our research.

We must also mention that some of the field and laboratory studies were carried out by past and present members of the Laboratory of Soil Science, Faculty of Agriculture, Kyoto University; without their devotion this research could not have been completed. We wish to acknowledge the help rendered by Prof. T. Hattori of Kyoto Prefectural University in the clay mineralogical studies.

The first draft of this book was kindly perused by Dr. Y. Ishizuka,

former Director of the Food and Fertilizer Technology Center at
Taipei, and by Dr. F. R. Moormann, Pedologist at the International
Institute for Tropical Agriculture in Ibadan, whose comments assisted
us greatly in the preparation of the final draft. We wish to express our
heartfelt thanks to them both.

We acknowledge with many thanks the assistance of Misses
U. Nakaoku and H. Hoshino in the preparation of the various drafts,
and of Mrs. S. Hawkes in improving our English.

May 15, 1977   Keizaburo Kawaguchi
Kyoto, Japan   Kazutake Kyuma

1

Introduction

The tropical Asia of our title corresponds to the broad geographic region of "tropical monsoon Asia," which comprises South and Southeast Asia. The countries of the Indian subcontinent—Pakistan, India, Nepal, Bhutan, and Bangladesh—together with Sri Lanka are usually called South Asia; and the area to the east of this and south of China, consisting of Burma, Thailand, Laos, Cambodia, Vietnam, Malaysia, Singapore, Indonesia and the Philippines, is called Southeast Asia.

These countries differ in many respects, race, language, religion, and so on, but have one important thing in common: rice cultivation. Of course, rice cultivation is not equally important to all these countries. Pakistan and the western states of India, in particular, have a climate which does not favor rice cultivation, but even Pakistan has about 1.5 million hectares of rice land and produces 3 million tons of paddy yearly, both of which exceed corresponding figures for Sri Lanka and West Malaysia.

The distinct characteristics of tropical Asia can best be seen from the basic figures given in Table 1.1. The table shows that tropical Asia

TABLE 1.1   BASIC STATISTICS FOR TROPICAL ASIA

POPULATION, LAND AND CROP	FIGURES FOR TROPICAL ASIA	PERCENT OF WORLD TOTAL
Population	$1{,}090{,}504 \times 10^3$	27.9
Land area	$897{,}060 \times 10^3$ ha	6.7
Arable land area	$262{,}112 \times 10^3$ ha	17.8
Ratio of arable to total land area	29.2 %	—
Rice { Acreage	$83{,}495 \times 10^3$ ha	61.0
Rice { Production	$151{,}356 \times 10^3$ ton	46.8
Cereals other than rice { Acreage	$79{,}796 \times 10^3$ ha	13.4
Cereals other than rice { Production	$65{,}677 \times 10^3$ ton	6.5

(Source: FAO Production Yearbook, 1974)

covers only about 7 percent of the world's land area, whereas more than a quarter of the world's population inhabits the region. This is one of the most densely populated areas of the world and will continue to be so in the foreseeable future.

Agricultural land in tropical Asia makes up about 18 percent of the world's agricultural land, and the ratio of agricultural to total land in the area is 29 percent, which is nearly three times the world average and is almost as high as that for western Europe. Even so, quite intensive land use is necessary to meet the demands of the large population. Rice cultivation seems to be adapted to this situation; more than 60 percent of the world rice acreage is found in this region alone. But the yield is not correspondingly high (see the yearly production figures).

Two points that emerge from Table 1.1 should be stressed, one being the intensive land use, and the other the importance placed on rice cultivation. These points are elaborated below, using relevant statistics.

Two estimates are available concerning potentially arable land in tropical Asia. One is given in *World Food Problem* compiled by the Science Advisory Committee to the American President (The White House, 1967). As shown in Table 1.2, Asia as a whole (except the USSR) has 620×10^6 hectares of potentially arable land, and more than 80 percent of this area is already under cultivation. The committee's conclusion is that, "In Asia, if we subtract the potentially arable land area in which water is so short that one 4-months growing season is

TABLE 1.2   PRESENT POPULATION AND CULTIVATED LAND ON EACH
CONTINENT, COMPARED WITH POTENTIALLY ARABLE LAND

CONTINENT	POPULATION (in 1965, millions)	AREAS IN MILLION HA.			CULTIVATED LAND PER PERSON	B/A (percent)
		Total	Potentially arable	Culti-vated		
			(A)	(B)		
Africa	310	2984	724	156	0.50	22
Asia	1855	2704	620	512	0.28	83
Oceania	14	812	152	16	1.14	1
Europe	445	472	172	152	0.34	88
North America	255	2084	460	236	0.93	51
South America	197	1732	672	76	0.39	11
U.S.S.R.	234	2208	352	224	0.96	64
TOTAL	3310	12996	3152	1372	0.41	44

(Source: World Food Problem, White House, 1967)

impossible, there is essentially no excess of potentially arable land over that actually cultivated.''

Within tropical Asia, however, there are differences from one sub-region to another. FAO's "Indicative World Plan," compiled in 1968, shows the result of a sample survey for South and Southeast Asia, Table 1.3. The figure for land presently cultivated, 84 percent of potentially arable land, in Asia corresponds with that cited in the preceding table (83 percent). The corresponding figure for South Asia, excluding Sri Lanka, is as high as 93 percent, while that for Southeast Asia and Sri Lanka is only 44 percent. Of the countries of Southeast Asia, the insular ones, especially Malaysia and Indonesia, are better off than the continental ones. Of the latter, Burma seems to have an exceptionally large area yet to be developed.

We have to consider what kinds of soil are left for development. A table is again cited from *World Food Problem*, Table 1.4, showing areas of different soil groups in the tropical zone of Asia. The region covered in the table is not the same as our definition of tropical Asia, but as a general comparison it suffices. The largest area of potentially arable land is found in "highly weathered and leached soils" and in "alluvial soils." The former are mostly latosols and red-yellow podzolic soils (Oxisols and Ultisols), occurring mainly on lateritized plateaus and high terraces, and have a low production potential. "Alluvial soils" include not only alluvial soils proper on deltas and floodplains, but also some immature soils such as low humic gley soils occurring on low terraces (Entisols and Inceptisols). The greater part of these alluvial

TABLE 1.3   ARABLE LAND AREA IN THE WORLD, ASIA AND FAR EAST

	POTENTIAL		PRESENT (1962)		PROPOSED (1985)	
	million ha	land area (percent)	million ha	potential (percent)	million ha	potential (percent)
Total (studied)	1145	26	512	45	600	53
Asia and the Far East*	252	47	211	84	223	89
South Asia (— Sri Lanka)	201	48	187	93	193	96
Southeast Asia and Sri Lanka	47	47	21	44	27	57
Far East	4	28	3	81	4	97

*Asia and the Far East here means Sri Lanka, China (Taiwan), India, Pakistan, Bangladesh, Philippines, Republic of Korea, Thailand, West Malaysia.
(Source: Indicative World Plan, FAO, 1968)

TABLE 1.4   AREAS OF DIFFERENT SOIL GROUPS IN THE
TROPICAL ZONE OF ASIA

SOIL	TOTAL	POTENTIALLY ARABLE ($\times$ 10^6 hectares)	GRAZING
Light-colored soils; base rich	80	32	16
Dark-colored soils; base rich	54	24	4
Moderately weathered and leached soils	84	54	18
Highly weathered and leached soils	488	108	264
Shallow soils and sands	113	12	40
Alluvial soils	168	114	20
TOTAL	987	344	362

(Source: World Food Problem, White House, 1967) .

soils have been utilized in South Asia and the continental part of Southeast Asia. Relatively large areas remain, however, along the coastlines of the insular part of Southeast Asia, notably in Borneo and Kalimantan and in Sumatra.

We predict that, these as yet uncultivated lands in "highly weathered and leached soil" and "alluvial soil" areas, will sooner or later be utilized for agriculture. But the former will pose great difficulties in management owing to the susceptibility of the soils to erosion and their low fertility, and many parts of the latter consist of problematic acid soils and peats. These points will be discussed in later chapters.

The importance of rice in the agriculture and national economy of tropical Asian countries is seen in Table 1.5. The proportion of rice acreage to total arable (area for annual crops) or agricultural (both arable land and land under permanent crops) land area ranges from 8.1 percent for Pakistan to 111.3 percent for Bangladesh. The value for Bangladesh is probably due to double-cropping.

Pakistan holds a peculiar position in tropical Asia with respect to rice cultivation. Her total rice acreage of 1.5 million hectares is completely irrigated, and rice is cultivated partially as a commercial crop. Pakistan's per capita production of cereals other than rice exceeds that of rice by a great margin, yet rice exports have earned quite a sizable amount of foreign exchange in recent years. Rice cultivation in northwestern India resembles Pakistan's with a low percentage of rice acreage, low per capita rice production, and high per capita output of cereals other than rice.

TABLE 1.5   IMPORTANCE OF RICE IN NATIONAL ECONOMY OF TROPICAL ASIAN COUNTRIES

COUNTRY	RICE ACREAGE (percent of Arable Land Area)	PER CAPITA RICE PRODUCTION[1] (in kg, 1974)	PER CAPITA OTHER CEREAL PRODUCTION (in kg, 1974)	IMPORT OR EXPORT[4] OF RICE[1] (× 1000 ton)		RICE TRADE[4] (percent Total External Trade)	
				1963–1965	1971–1973	1963–1965	1971–1973
Bangladesh	111.3	129.7	2.1	—	−497	—	−5.7
Burma	51.4[3]	180.2	5.5	+1485	+485	+61.0	+32.3
Cambodia	38.5	51.4	8.7	+447	−56	+52.1	−13.7
India	23.3	68.2	79.6	−634	−360	−2.6	−2.9
Indonesia	47.0	108.7	20.2	−765	−967	—	−13.5
W. Malaysia	97.9	115.8	0.6	−355	−135	−5.4	−1.4
Pakistan	8.1[2]	30.2	128.1	+78	+370	+1.5	+8.6
Philippines	40.1	83.1	52.3	−375	−385	−6.0	−2.6
Sri Lanka	71.3	87.8	2.7	−568	−316	−16.8	−9.4
Thailand	62.2	207.8	65.0	+1786	+1514	+34.7	+15.9
Vietnam Rep.	92.2	240.7	8.3	−35	−227	+11.9	−9.5

1   As milled rice (paddy weight × 0.65)
2   Based on agricultural area
3   Based on total cultivated land area (including fallow land), cited from the Government of Burma Publication, 1972
4   Export is indicated by + ; import by − .
(Source: FAO Production Yearbook, 1974; FAO Trade Yearbook, 1969, 1974)

The other countries, including the eastern part of India, are more heavily dependent on rice as the staple food. Burma and Thailand are the rice exporters, and the others import rice. Cambodia and (South) Vietnam need special mention. They had been rice exporters before the recent war. The drastic decrease of rice acreage in Cambodia (see Table 1.6) during the last ten years, and the rice imports of Vietnam in spite of her large per capita output of rice clearly show the impact of the war. In our view these two countries will be able to return to self-sufficiency and may even have some exportable surpluses of rice after their recovery from the ravages of war.

Burma's export capacity has declined sharply during the last ten years from 1.5 million tons to a few hundred thousand tons, mainly because of stagnation in production (see Table 1.6) and an increase in domestic consumption. Thailand exports 1 to 1.5 million tons of rice annually, earning a considerable amount of foreign exchange. She exports a large quantity of maize and this, in part, explains the decreased contribution of rice to the value of total exports in the recent years.

Of the rice importers, Indonesia and the Philippines have relatively large areas that produce cereals other than rice. In these two countries there are areas where people depend on maize as their staple food. But even these people are potential rice consumers, and the need for rice will become more acute as the national income rises. West Malaysia and Sri Lanka once depended very heavily on rice imports, but their efforts to reduce the dependence in the last ten years have met with remarkable success.

If we consider the recent rapid increases in population, the importance of rice in most countries of tropical Asia will remain the same or become even greater in the coming years. Population increase in most countries is estimated to range from 2.5 to 3 percent, which results in a doubling of population in twenty-three to thirty years. As shown in Table 1.6, the increase in rice production in the period from 1961–1965 to 1972–1974 greatly exceeds that of population increase in such countries as Indonesia, West Malaysia, Pakistan, and Sri Lanka, but is not as great as population increase in some other countries, notably, Burma, Bangladesh, India, and Thailand. (Cambodia and South Vietnam have been excluded from consideration.) The great production increase in Pakistan, Indonesia, and Sri Lanka appears to have been achieved mainly through increase in yields, and this is most probably a result of recent technological innovations, part of the green revolution. Even in India and Burma, though overall production increase is not remarkable, the effect of yield increase outweighs the effect of area increase.

Table 1.6 Recent Changes in Rice Cultivation in Tropical Asian Countries

Country	Population Increase	Production (× 1000 ton)			Area (× 1000 ha)			Yield (kg/ha)			Contribution to Production Increase	
	1963–1973 (percent)	1961–1965 mean	1972–1974 mean	Increase (percent)	1961–1965 mean	1972–1974 mean	Increase (percent)	1961–1965 mean	1972–1974 mean	Increase (percent)	Area (percent)	Yield (percent)
Bangladesh	41.2	15034	17237	14.7	8955	9721	8.6	1679	1774	5.7	60	40
Burma	25.8	7786	8122	4.3	4741	4833	1.9	1642	1680	2.5	45	55
Cambodia	33.5	2461	1204	−51.1	2284	989	−56.7	1077	1217	13.0	−111	11
India	24.3	52733	61994	17.6	35626	37400	5.0	1480	1657	12.0	30	70
Indonesia	31.8	12393	20910	68.7	7036	8316	18.2	1761	2514	42.8	32	68
W. Malaysia	30.1	957	1702	77.8	382	578	51.3	2503	2943	17.6	72	28
Pakistan	35.3	1824	3450	89.1	1287	1505	16.9	1417	2296	62.0	25	75
Philippines	39.4	3957	5362	35.5	3147	3402	8.1	1257	1575	25.3	26	74
Sri Lanka	27.4	967	1500	55.1	505	584	15.6	1914	2552	33.3	33	67
Thailand	38.1	11267	13495	19.8	6348	7425	17.0	1775	1818	2.4	87	13
Vietnam Rep.	24.3	5029	6858	36.4	2472	2810	13.7	2034	2439	19.9	41	59

(Source: FAO Production Yearbook, 1974)

In other countries, area increase seems to have contributed relatively more to the increase in production. In West Malaysia double-cropping made an important contribution to areal expansion, while in Thailand reclamation of marginal lands seems to have resulted in an increase in area, which would explain the relatively low increase in yield for the country as a whole. The latter type of inefficient areal expansion ought to be avoided in the interests of the soil and watershed conservation.

Thus, there are two ways to increase production of rice without increasing the area cultivated. One is the reinforcement of land productivity through technological innovation, and the other is a further intensification of land use through double or multiple cropping. Although the initial enthusiasm has died away, the green revolution was welcome because it pointed in the right direction. The Committee for World Food Problems (The White House, 1967) also concluded that, "To increase food supplies in Asia, therefore, it will be necessary either to increase yields per acre or to increase the gross harvested area through double or triple cropping. . . ."

Multiple cropping would also enable the countries of tropical Asia to pursue another line of agricultural development, that is, crop diversification. Crop diversification would be an adaptation to the change in the economic environment of the region; the countries of tropical Asia have to produce such agricultural commodities as beans and maize for export, and sugarcane and cotton to meet the increasing domestic demand. Because most countries in tropical Asia cannot reduce their rice acreage, crop diversification can be achieved either by reclamation of upland areas or by multiple cropping on the existing arable land.

Both intensification and diversification processes should be undertaken so as not to disrupt the equilibrium that has been established between the traditional pattern of land use and the natural environment. The failure of some of the modernization and development programs in agriculture was often caused by hasty transplantation of modern technology without thorough understanding of the existing conditions. Knowledge of soil conditions is one of the prerequisites in such cases of technology transfer.

And so the prime objective of this book is to contribute to a better understanding of paddy soils, which are the most important medium of agricultural development in tropical Asia. Furthermore, a study of paddy soils is mainly a study of relatively intensively cultivated soils on alluvial lowlands, for which no established method of capability evaluation has yet been made available. Here we wish to give an example of an approach to this problem.

First, we attempt to list the reasons rice cultivation has been and is the main form of agriculture in tropical Asia, and for this, in chapter 2, we deal with general climatic and physiographic features of tropical Asia's paddy lands.

Before the detailed discussion, general pedological and edaphological features of paddy soils in tropical Asia are presented in chapter 3.

As there have been relatively few studies on paddy soils in the countries of tropical Asia, we conducted a comparative study of paddy soils in many countries within the region ourselves. The methods adopted in the study and a general description of the samples used are given in chapter 4.

The range of variation of ordinary chemical, total chemical, mechanical and clay mineralogical properties of paddy soils is presented for tropical Asia as a whole, and for each country within the region, on the basis of our studies in chapter 5. The interrelationships between these properties are also discussed.

We hold the view that the nature of parent materials is of paramount importance in determining the fertility status of paddy soils, and a classification of paddy soil materials in tropical Asia is attempted in chapter 6.

Fertility status of paddy soils in tropical Asia will be numerically assessed by a unique method, presented, with results, in chapter 7.

A little more detail of material and fertility characteristics of the samples from each country will be given in chapter 8, so that regional differences and problem areas within each country may be clarified.

Special problems related to paddy soils in tropical Asia, such as silting from floods, acid sulfate soils, nutrient deficiency, and toxicity are discussed in chapter 9.

In our conclusion, chapter 10, we consider the role of soils study in the intensification and diversification of agriculture in tropical Asia.

2

Environmental Conditions of Rice Cultivation in Tropical Asia

The traditional agriculture of a region reflects the pattern of man's adaptation to the given natural environment. Climate, physiography, water and soil conditions determine, more or less, the type of crops to be cultivated and the management practices to be adopted. Of course, the pattern of adaptation is also determined by the level of technology that is available to man for modifying the elements of the natural environments.

Rice cultivation in tropical Asia is also thought to represent the pattern of adaptation which has been sought by the people in the region since the time agriculture began. As climate and physiography are the two main constraints for rice cultivation in their combined effect on water conditions, we first discuss these two factors in relation to rice cultivation.

2.1 CLIMATE

Tropical Asia is situated in the southernmost part of "monsoon Asia," and its climate can be broadly categorized as tropical monsoon climate. Monsoon means "seasonal wind," and the monsoon climate is characterized by a change in wind direction according to the seasons. Winds which blow across the ocean bring moisture to the land as monsoonal rains. Thus, a change in wind direction often results in alternation of wet and dry seasons. Orographic conditions considerably modify this basic pattern and produce many climatic variations within the region. Therefore, we must consider the climatic regions of tropical Asia.

Climatic Regions

There have been many attempts to establish climatic regions. Köppen's system is most well known, dividing tropical Asia into three main climatic regions: (1) tropical rainy climate (Af), (2) tropical wet monsoon climate (Am), and (3) tropical climate with wet and dry seasons or tropical savanna climate (Aw). Roughly speaking, the tropical rainy climate or Af occurs in lowlands within 5–10° of the equator, the tropical climate with wet and dry seasons or Aw occurs north and south of the Af zone, and the tropical wet monsoon climate or Am is a transitional type (Money, 1965).

The method proposed by Thornthwaite (1948) also is widely used and has unique features in that water need is computed from mean monthly temperature as potential evapotranspiration, and soil moisture retention is taken into consideration in assessing water surplus and deficiency. As "soil forming processes are related to water surpluses and deficiencies" (Thornthwaite, 1948), this method has been preferred to other classification schemes by many soil scientists. We, too, adopted Thornthwaite's method in studying the water balance in tropical Asia (Kyuma, 1971).

Recently we examined the applicability of a numerical taxonomic method to the climatic classification of Japan, with the intention of establishing climatic regions as a basis for classification of alluvial soils (Kyuma, 1972a). As noted in this study, the numerical method gave readily mappable climatic regions with a fairly good internal coherence, while Thornthwaite's method, used for comparison, presented difficulty in mapping. Therefore, we have used the numerical method for establishing climatic regions in tropical Asia (Kyuma, 1972b).

One hundred and twenty-five stations, of some 300 stations for which mean monthly temperature and rainfall data are available, were chosen, taking density of distribution of stations and variability of climate in an area into consideration. The distribution of the stations is shown in Figure 2.1. The fundamental limitation in the selection of sample stations was that there are many areas for which meteorological data are very scarce, for example, Laos and the Indonesian islands other than Java. Another consideration in selecting the stations was to exclude those on high land. This was because the major rice-growing lands are in the lowlands. Location, moisture data, and climatic type, by Thornthwaite's method for the sample stations, are given in Table 2.1.

Based on the mean monthly temperature and mean monthly rainfall data for the 125 stations, 9 climatic groups were established by means of numerical taxonomic procedures after Sokal and Sneath (1963) (see chapter 6). The mean vectors for each of the 9 groups are illustrated in Figure 2.2. Although the distribution of stations belonging to each of the groups showed fairly good regionality, one difficulty in delineating climatic regions was the scarcity of the number of stations for each group relative to the area to be covered. In order to overcome

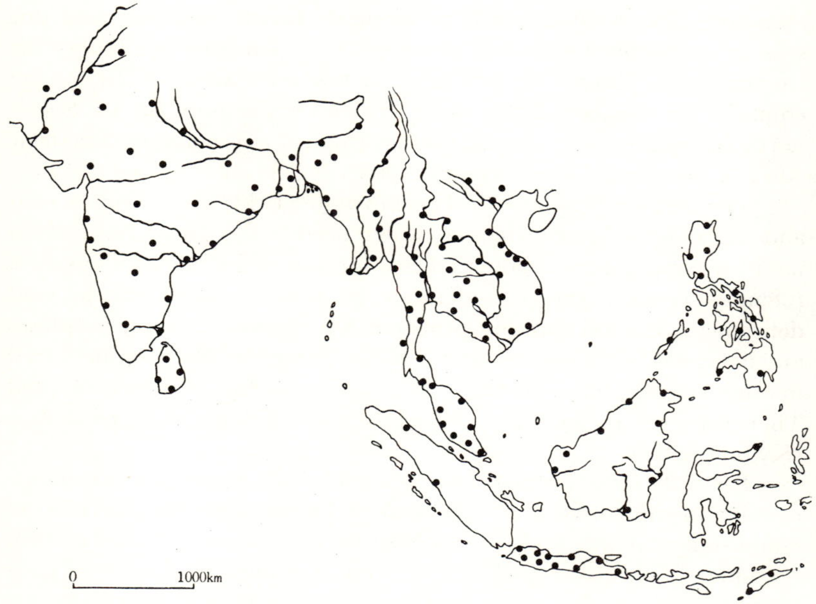

FIG. 2.1   MAP SHOWING DISTRIBUTION OF THE SAMPLE STATIONS

this difficulty we derived discriminant functions (see chapter 6) for all possible pairs of the groups. As these functions proved satisfactory (misclassification cases, 5 percent), the remaining 180 stations were classified by the discriminant functions and the results plotted on a map. Except for the Philippines, stations belonging to one group were so well clustered geographically that delineation of a climatic region as the mappable equivalent of the group was quite easy. Thus, the following 9 climatic regions, which correspond to the 9 groups, are delineated on the map shown in Fig. 2.3.

Group I	Strait-Sunda Region
Group II	Malay-Northern Borneo Region
Group III	Oceanic Sumatra-West Java Region
Group IV	Southwest-Facing Coastal Region
Group V	Southern Indochina-Southern India Region
Group VI	Middle Vietnam Region
Group VII	Central India-Northern Indochina Region
Group VIII	Tongking-Assam Region
Group IX	Lower Indus Region

The region name are the typical areas with that climate. A brief description of each group or region is given as follows:

Region I   Humid equatorial climate with very small temperature fluctuation; monthly rainfall ranges from about 100 mm to 250 mm, with the minimum in the summer season.

Region II   Humid to perhumid equatorial climate with very small temperature fluctuation; monthly rainfall fluctuates between 150 and 200 mm for the greater part of a year, but it attains a sharp maximum of 350–400 mm during October–December under the influence of the northeast monsoon.

Region III   Perhumid equatorial climate with very small temperature fluctuation; monthly rainfall exceeds 200 mm throughout the year.

Region IV   Tropical monsoon climate with small temperature fluctuation; monthly rainfall changes drastically from less than 10 mm during the winter months to more than 500 mm during the summer months.

Region V   Tropical monsoon climate with small temperature fluctuation; monthly rainfall does not exceed 200 mm even during the monsoon season and it is quite low during the winter season.

Region VI   Tropical monsoon climate with moderate temperature fluctuation; typically under the influence of the northeast monsoon, monthly rainfall reaches a maximum of about 500 mm during October–November and a minimum of about 50 mm in March–April.

Region VII   Tropical to subtropical monsoon climate; severe drought during the winter season and moderate rainfall of 250–300 mm during July–September.

Region VIII   Subtropical monsoon climate; winter drought is not severe, with a minimum monthly rainfall of 30 mm, and summer rainfall of more than 200 mm lasts from May to September.

Region IX   Subtropical arid climate; slightly monsoonal with a rainfall maximum in July–August of 60–70 mm, but for the rest of the year rainfall does not exceed 30 mm.

Taking these climatic regions as the frame of reference, we now consider two important climatic factors in relation to rice cultivation, water regime and solar radiation.

Soil Water Regime

Water conditions in rice lands may be broadly grouped into three categories: rain-fed, irrigated, and water conserved. The latter is important in the relative depressions of alluvial plains and the deltas of great rivers where water accumulates. Thus, in water conservation areas physiographic factors overshadow climatic factors in influencing water conditions.

Irrigated areas have more or less elaborate irrigation installations, such as weirs and canals or sometimes tube wells with pumps, to secure

TABLE 2.1 LIST OF METEOROLOGICAL STATIONS, THEIR LOCATION AND THORNTHWAITE'S CLIMATIC TYPE

STATION No.	Location	Latitude	Altitude m	Water need (PE or TE)	Precipi-tation	Water surplus	Water deficiency	Moisture index (Im)	Climatic type	Climatic region
1	Laokay, Vietnam	22°30′N	103	1231	1750	519	0	42.2	$B_2A'r$	VIII
3	Langson, ,,	21°51′N	265	1116	1428	312	0	28.0	$B_1B_4'r$	VIII
6	Hanoi, ,,	21°30′N	8	1256	1673	423	5	33.4	$B_1A'r$	VIII
8	Dong Hoi, ,,	17°29′N	20	1388	2118	1059	331	62.0	$B_3A's$	VI
9	Vinh, ,,	18°39′N	6	1286	1828	630	88	44.9	$B_2A'r$	VIII
10	Hue, ,,	16°24′N	16	1351	2904	1790	237	122.0	$AA's$	VI
11	Quang Tri, ,,	16°44′N	7	1423	2541	1445	326	87.8	$B_4A's$	VI
12	Phanthiet, ,,	10°56′N	6	1623	1216	130	537	−11.8	$C_1A'd$	V
13	Tourane, ,,	16° N	8	1435	1974	1027	487	51.2	$B_2A's_2$	VI
15	Qui Nhou, ,,	13°45′N	6	1567	1653	711	625	21.4	$B_1A's_2$	VI
21	Saigon, ,,	10°47′N	11	1685	1806	560	440	17.6	$C_2A'w$	V
23	Hatien, ,,	10°10′N	3	1689	1958	572	302	23.1	$B_1A'w$	V
24	Stung Treng, Cambodia	13°31′N	54	1625	1729	565	460	17.8	$C_2A'w$	V
25	Siemreap, ,,	13°22′N	15	1616	1281	202	537	−7.4	$C_1A'w$	V
29	Phnom Penh, ,,	11°33′N	12	1750	1320	92	523	−12.7	$C_1A'd$	V
30	Luang Prabang, Laos	19°53′N	304	1453	1203	50	300	−8.9	$C_1A'd$	VII
32	Vientiane, ,,	17°57′N	170	1535	1714	578	400	22.0	$B_1A'w$	VII
33	Pakse, ,,	15°07′N	96	1673	1564	526	635	8.7	$C_2A'w_2$	V
37	Kota Bharu, W. Malaysia	6°08′N	6	1688	2755	1138	72	63.4	$B_3A'r$	II
40	Kuala Lumpur, ,,	3°07′N	39	1720	2499	779	0	45.3	$B_2A'r$	II
42	Singapore, ,,	1°18′N	10	1716	2232	516	0	30.1	$B_1A'r$	I
247	Kulim, ,,	5°23′N	32	1730	3128	1399	0	80.9	$B_4A'r$	II
249	Sitiawan, ,,	4°13′N	7	1710	1902	296	106	13.6	$C_2A'r$	I
251	Kuantan, ,,	3°46′N	19	1589	3003	1414	0	89.0	$B_4A'r$	II
255	Segamat, ,,	2°30′N	29	1742	1991	258	10	14.5	$C_2A'r$	I
44	Sandakan, E. Malaysia	5°50′N	46	1773	3137	1364	0	76.9	$B_3A'r$	II
46	Labuan, ,,	5°17′N	18	1772	3573	1803	0	101.8	$AA'r$	II
49	Bintula, ,,	3°11′N	3	1587	3499	1914	0	120.6	$AA'r$	II

No.	Station	Alt. (m)	Lat.	(a)	(b)	(c)	(d)	(e)	Climate	Zone
50	Kuching, ″	26	1°29′N	1693	3905	2214	0	130.8	AA′r	III
51	Tarakan, Indonesia	12	3°19′N	1613	3869	2257	0	140.0	AA′r	III
52	Balikpapan, ″	7	1°17′N	1579	2228	649	0	41.1	$B_2A′r$	I
53	Pontianak, ″	3	0°01′N	1698	3178	1481	0	87.2	$B_4A′r$	II
54	Mapanget, ″	86	1°32′N	1395	3398	2002	0	143.5	AA′r	III
57	Dili, ″	0	8°35′S	1785	848	0	−938	−31.5	DA′d	I
60	Kupang, ″	45	10°10′S	1647	1440	542	749	5.6	$C_2A′s_2$	I
62	Medan, ″	23	3°35′N	1516	2029	512	0	33.8	$B_1A′r$	I
64	Padang, ″	7	0°56′S	1626	4172	2547	0	156.6	AA′r	III
67	Surabaya, ″	7	7°16′S	1670	1674	439	436	10.6	$C_2A′s$	I
71	Jember, ″	83	8°09′S	1461	1868	595	187	33.0	$B_1A′r$	I
74	Semarang, ″	10	7°00′S	1643	2033	519	129	26.6	$B_1A′r$	I
77	Wedi-Birit, ″	150	7°45′S	1519	1846	582	254	28.3	$B_1A′s$	I
80	Kuyper, ″	2	6°02′S	1791	1603	276	464	−0.1	$C_1A′s$	I
87	Buitenzorg (Bogor), ″	250	6°35′S	1385	3934	2550	0	184.1	AA′r	III
231	Banjarmasin, ″	2	3°19′S	1678	2755	1077	0	64.2	$B_3A′r$	I
232	Cirebon, ″	4	6°42′S	1809	2209	723	324	29.2	$B_1A′s$	I
233	Cilatjap, ″	6	7°44′S	1614	4274	2659	0	164.8	AA′r	III
234	Subang, ″	95	6°35′S	1539	3369	1855	26	119.5	AA′r	III
245	Bandung, ″	768	6°55′S	1100	2159	1060	0	96.4	$(B_4B′_4r)$	III
88	Aparri, Philippines	4	18°22′N	1657	2307	916	266	45.6	$B_2A′r$	VI
90	Echague, ″	78	16°42′N	1667	1583	279	365	3.6	$C_2A′s$	V
92	Tacloban, ″	21	11°15′N	1737	2210	472	0	27.2	$B_1A′r$	I
93	Cebu City, ″	42	10°20′N	1771	1789	213	194	5.5	$C_2A′r$	V
94	Manila Airport, ″	15	14°31′N	1715	1791	572	497	16.0	$C_2A′w$	IV
95	Legaspi City, ″	19	13°08′N	1693	3407	1713	0	101.2	AA′r	II
96	Cuyo, ″	4	10°51′N	1778	2228	939	491	36.2	$B_1A′w$	IV
98	Iwahig, ″	14	9°44′N	1652	1963	556	244	24.8	$B_1A′r$	I
100	Dagupan City, ″	2	16°03′N	1821	2002	731	552	22.0	$B_1A′w$	IV
101	Zamboanga City, ″	6	6°58′N	1703	1302	0	401	−14.1	$C_1A′d$	V
103	Davao, ″	20	7°04′N	1689	1971	282	0	16.7	$C_2A′r$	I
104	Bhamo, Burma	118	24°16′N	1363	1857	692	199	42.0	$B_2A′r$	VII
106	Mandalay, ″	76	21°59′N	1626	776	0	850	−31.4	DA′d	V
107	Akyab, ″	5	20°08′N	1491	4778	3606	318	229.1	AA′w	IV
108	Yamethin, ″	119	20°25′N	1581	966	0	616	−23.4	DA′d	V

TABLE 2.1 CONTINUED

No.	STATION / LOCATION	LATITUDE	ALTITUDE m	WATER NEED (PE or TE)	PRECIPI-TATION	WATER SURPLUS	WATER DEFICIENCY	MOISTURE INDEX (Im)	CLIMATIC TYPE	CLIMATIC REGION
109	Toungoo, Burma	18°55′N	48	1564	2105	947	406	45.0	$B_2A'w$	IV
110	Victoria Point, ,,	9°58′N	47	1720	3964	2653	408	140.0	$AA'w$	IV
111	Rangoon, ,,	16°46′N	23	1693	2529	1409	572	62.9	$B_3A'w_2$	IV
112	Diamond Island, ,,	15°51′N	12	1686	3117	1915	483	96.4	$B_4A'w$	IV
113	Amherst, ,,	16°05′N	22	1655	5156	3983	483	223.2	$AA'w$	IV
115	Mergui, ,,	12°26′N	20	1596	4123	2796	270	165.0	$AA'w$	IV
118	Srimangai, Bangladesh	24°19′N	21	1408	2509	1116	13	78.7	$B_3A'r$	VIII
120	Jessore, ,,	23°10′N	7	1500	1608	309	200	12.6	$C_2A'r$	VII
122	Cox's Bazar, ,,	21°26′N	4	1440	3560	2261	142	151.1	$AA'r$	IV
124	Bogra, ,,	24°51′N	20	1484	1776	539	247	26.3	$B_1A'r$	VII
286	Lahore, Pakistan	31°33′N	214	1396	492	0	−9036	−38.8	DA'd	IX
292	Multan, ,,	30°12′N	123	1479	167	0	−13117	−53.2	EA'd	IX
294	Khanpur, ,,	28°39′N	91	1468	165	0	−13025	−53.3	EA'd	IX
308	Jacobabad, ,,	28°18′N	56	1537	990	0	−14379	−56.1	EA'd	IX
310	Hyderabad, ,,	25°23′N	29	1616	157	0	−14589	−54.2	EA'd	IX
127	Mohanbari, Assam, India	27°29′N	111	1229	2775	1548	0	126.0	$AA'r$	VIII
128	Gauhati, ,,	26°05′N	54	1329	1634	395	91	25.6	$B_1A'r$	VIII
130	Silchar, ,,	24°49′N	29	1423	3263	1849	8	129.6	$AA'r$	VIII
136	Calcutta, W. Bengal, India	22°32′N	6	1609	1582	365	393	8.0	$C_2A'w$	V
138	Cuttack, Orissa, India	20°48′N	27	1675	1545	412	543	5.1	$C_2A'w$	V
140	Darbhanga, Bihar, India	26°10′N	49	1458	1258	251	449	−1.3	$C_1A'w$	VII
145	Daltonganj, ,,	24°03′N	221	1448	1237	325	537	0.2	$C_2A'w_2$	VII
146	Jamshedput, ,,	22°49′N	129	1545	1442	360	463	5.3	$C_2A'w$	VII
148	Kanpur, U. P., India	26°26′N	126	1470	883	84	672	−21.7	DA'd	VII
154	Raipur, M. P., India	21°14′N	298	1548	1360	442	630	4.1	$C_2A'w_2$	VII
265	Sagar, ,,	23°51′N	551	1401	1394	6033	6098	17.0	$C_2A'w_2$	VII
158	Masulipatnam, A. P., India	16°11′N	7	1697	1077	33	652	−21.1	DA'd	V
159	Hyderabad, ,,	17°27′N	545	1455	761	0	695	−28.7	DA'd	VII

160	Visakhapatnam, ,,	17°43′N	3	1734	944	21	811	−26.9	DA'd	V
162	Madras, Madras, India	13°00′N	16	1789	1233	275	829	−12.4	C_1A's	V
163	Nagapatlinam, ,,	10°46′N	9	1798	1367	450	881	−4.4	C_1A's_2	V
166	Coimbatore, ,,	11°00′N	409	1581	614	0	968	−36.7	DA'd	V
167	Belgaum, Mysore, India	15°51′N	753	1257	1492	701	465	33.6	B_1A'w_2	VII
168	Mangalore, ,,	12°52′N	22	1709	3479	2367	596	117.6	AA'w_2	IV
169	Bellary, ,,	15°09′N	449	1705	520	1185	0	69.5	B_3A'r	V
152	New Delhi, India	28°35′N	216	1446	715	0	730	−30.3	DA'd	VII
267	Kotah, Rajasthan, India	25°11′N	—	157	85	119	842	−24.6	DA'd	VII
268	Bikaner, ,,	28°00′N	—	148	31	0	−1179	−47.7	EA'd	IX
270	Ahmedabad, Gujarat, India	23°04′N	—	154	38	0	−116	−45.2	EA'd	VII
275	Bombay (Calaba), Maharastra, India	18°54′N	11	1708	2078	12064	8368	41.2	B_2A'w_2	IV
276	Ratnagiri, ,,	16°59′N	35	1642	2617	17517	−7769	78.3	B_3A'w_2	IV
281	Akola, ,,	20°42′N	282	1602	877	426	−7674	−26.1	DA'd	VII
170	Colombo, Sri Lanka	6°54′N	6	1690	2397	706	0	41.8	B_2A'r	II
173	Hambantota, ,,	6°07′N	20	1709	1074	0	637	−22.4	DA'd	V
180	Anuradhapura, ,,	8°21′N	89	1689	1446	289	531	−1.8	C_1A's	V
184	Batticaloa, ,,	7°43′N	12	1716	1756	702	661	17.8	C_2A's_2	II
188	Chiang Rai, Thailand	19°55′N	378	1393	1747	589	233	32.3	B_1A'w	VII
190	Chiang Mai, ,,	18°47′N	314	1507	1249	141	399	−6.5	C_1A'd	VII
197	Tak, ,,	16°50′N	—	1678	951	0	727	−26.0	DA'd	V
198	Loei, ,,	17°32′N	—	1487	1194	39	331	−10.7	C_1A'd	VII
202	Mukdahan, ,,	16°33′N	138	1554	1447	338	445	4.6	C_2A'w	VII
204	Chaiyaphum, ,,	15°45′N	—	1634	1089	31	577	−19.3	C_1A'd	V
207	Surin, ,,	14°53′N	145	1654	1341	134	448	−8.1	C_1A'd	V
210	Nakhon Sawan, ,,	15°48′N	28	1794	1182	33	644	−19.7	C_1A'd	V
213	Kanchanaburi, ,,	14°01′N	28	1741	992	0	749	−25.8	DA'd	V
215	Bangkok, ,,	13°44′N	—	1806	1409	184	522	−7.2	C_1A'w	V
217	Aranyaprathet, ,,	13°42′N	44	1738	1522	177	394	−3.4	C_1A'w	V
219	Chanthaburi, ,,	12°37′N	5	1719	3027	1647	338	84.0	B_4A'w	IV
222	Prachuap Khiri Khan, ,,	11°48′N	5	1681	1163	54	572	−17.2	C_1A'd	V
224	Ban Dan, ,,	9°08′N	3	1725	1858	384	251	13.5	C_2A'r	V
226	Songkhla, ,,	7°13′N	4	1789	2232	858	416	34.0	B_1A's	II
230	Trang, ,,	7°30′N	—	1755	2178	661	238	29.5	B_1A'r	V

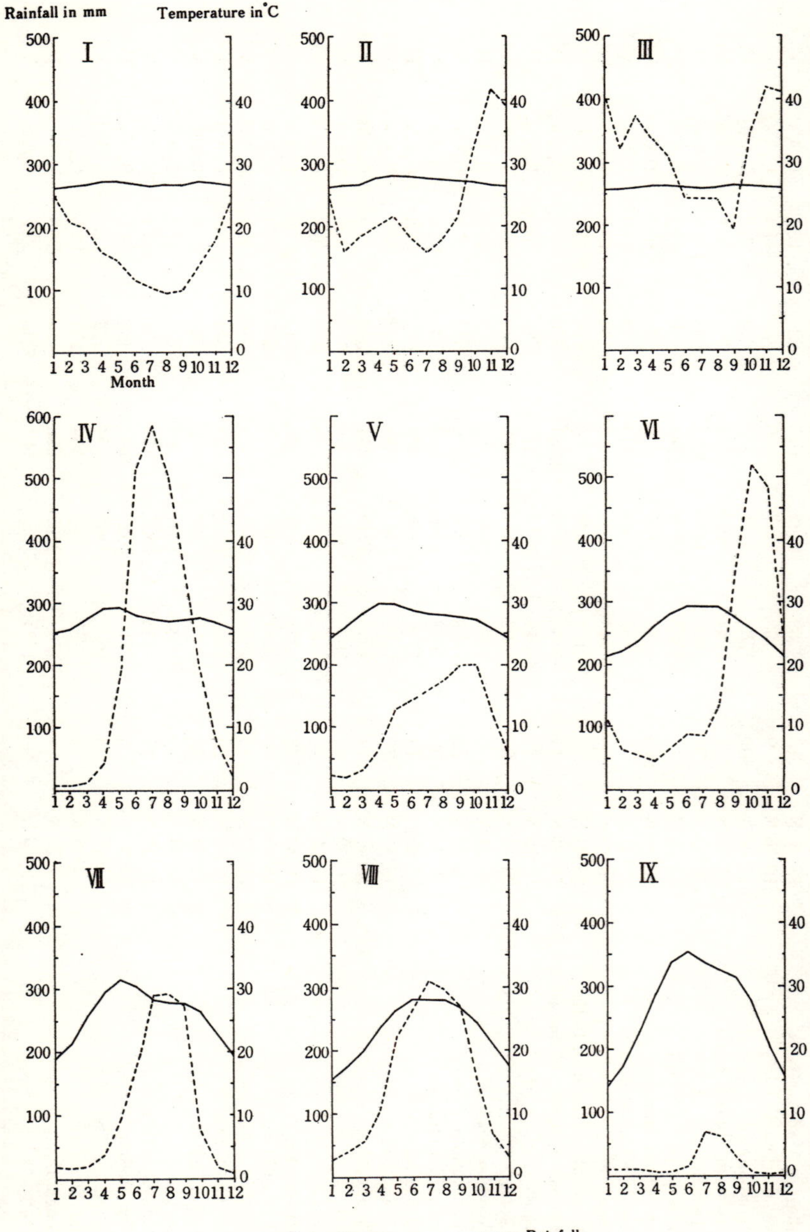

FIG. 2.2   SAMPLE MEAN VECTORS FOR THE NINE CLIMATIC REGIONS

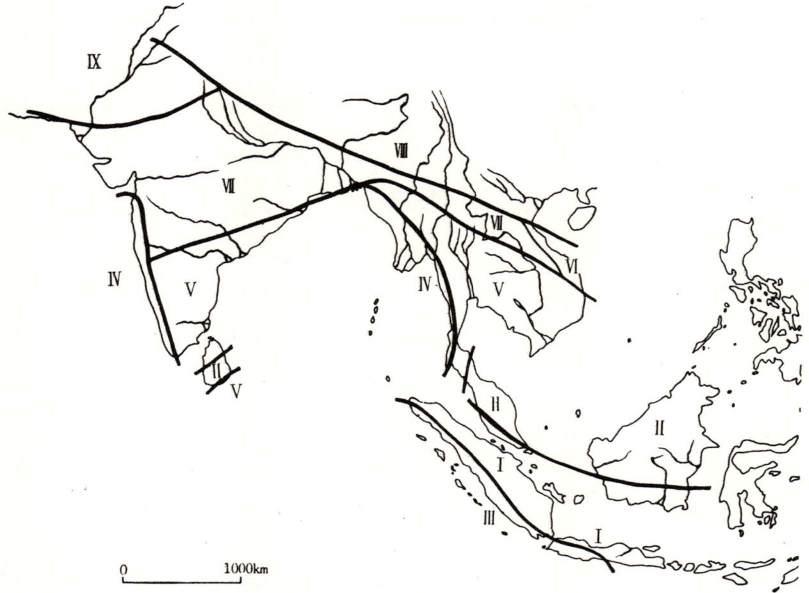

FIG. 2.3　MAP SHOWING CLIMATIC REGIONS

and distribute water effectively. This applies not only to government administered irrigation systems but also to communally operated ones, though these systems may be quite different in scale.

Even so-called rain-fed rice may have some means of irrigation, such as small reservoirs, ditches, and dikes. However, the source of water is localized and limited, and the influence of climate and such factors as gradient and catchment-rice area ratio have a direct bearing on water conditions in rain-fed rice lands. Thus, rice production here is controlled to a great extent by nature's whims.

Unfortunately pertinent statistical data on areas under different water conditions are not available. Table 2.2 shows the area of irrigated land and total area of rice land in eight tropical Asian countries in recent years. Two points should be noted in relation to the assessment of the extent of irrigated rice lands from the data. One is that the irrigated area includes the area of irrigated nonrice lands, such as those in Central and East Java under sugarcane cultivation. The other is that the total rice land area quoted is the harvested area, in which a higher percentage of irrigated land occurs than in planted-but-not-harvested area. Even with these positive errors, the area of irrigated rice land

TABLE 2.2   AREAS OF IRRIGATED LAND AND TOTAL RICE LAND
IN TROPICAL ASIAN COUNTRIES*

Country	Rice Land Area (× 10³ ha)	Irrigated Area (× 10³ ha)	Ratio (percent)
Bangladesh	9904	1212	12.2
Burma	5059	839	16.6
Cambodia	681	89	13.1
Indonesia	8500	6900	81.2
West Malaysia	597	246	41.2
Sri Lanka	638	465	72.9
Thailand	7734	2476	32.0
Vietnam Rep.	2900	580	20.0
Total	36013	12807	35.6

*Pakistan and India are excluded because of their large acreages of irrigated nonrice land.
(Source: FAO Production Yearbook, 1974)

constitutes only less than 40 percent of the area of rice land for the countries listed in the table, and the remainder is still in rain-fed conditions.

The following is an attempt to clarify the soil water regime of rice lands in tropical Asia, using the available climatic data. As the approach taken herein has many obvious limitations, the applicability of the results is also limited. The findings in this study are more directly applicable to the rain-fed and irrigated rice lands than to the water conservation areas.

The stations selected for the study of soil water regime are same 125 stations listed in Table 2.1 that were used for establishing climatic regions. In the same table, water need or potential evapotranspiration estimated by Thornthwaite's method is given for each of the stations.

The method of water regime computation is basically the same as that used in Thornthwaite's scheme except for the following consideration: levees or banks are built around each rice field in order to pond water during the cropping season. Thus, the following assumptions were put forward for computing soil water regimes:

1. The soil can retain 100 mm of water in the effective rooting depth of the soil profile, as Thornthwaite assumed.

2. The bank height is 20 cm; a maximum of 200 mm of water can be retained and ponded on the field and any excess is lost as run-off.

3. Percolation loss from the ponded water amounts to a maximum

TABLE 2.3   COMPUTATION OF SOIL WATER REGIME
FOR STATION 111 (RANGOON) (in mm.)

MONTH	1	2	3	4	5	6	7	8	9	10	11	12	ANNUAL
Potential Evapotranspiration	92	100	145	167	176	158	162	152	147	152	138	103	1692
Rainfall	8	5	6	17	260	524	492	574	398	208	34	3	2529
Soil Moisture	0	0	0	0	84	100	100	100	100	100	100	0	—
Ponded Water	0	0	0	0	0	200	200	200	200	156	0	0	—
Percolation	0	0	0	0	0	100	100	100	100	100	52	0	552
Run-off	0	0	0	0	0	50	230	322	151	0	0	0	753
Deficiency	84	95	139	150	0	0	0	0	0	0	0	0	468

of 100 mm per month, which corresponds to the hydraulic conductivity of about 4×10^{-6} cm/sec.

From the difference between actual precipitation and potential evapotranspiration, which was computed according to Thornthwaite's method, the surplus or deficiency of water was computed for each month, allowing for the above assumptions. An example of computation is given in Table 2.3.

The results are summarized in Tables 2.4 and 2.5, and some examples of water regime patterns are illustrated in Figure 2.4 for stations selected from each of the climatic regions established earlier.

In Table 2.4 the stations are grouped according to the available water regime, based on the number of consecutive months over which soil moisture is available. It is obvious from the table that climatic regions I, II, III, IV, VI, and VIII have more favorable water regimes than climatic regions V, VII, and IX. This is not surprising for Region IX, which is classified either as arid or semiarid by Thornthwaite's method, but it is rather surprising to find that a fairly high percentage of the area within Regions V and VII has a water regime so severe that it does not seem to be able to support even a single short-term upland crop.

Table 2.5 shows the number of stations in each climatic region in each category of ponded water regimes. In the preparation of this table, a water regime that has two or less consecutive months of ponded water is thought to be prohibitive of rice cultivation. A regime having three consecutive months of ponded water is regarded as marginal. Four to six months of ponded water may be sufficient for one crop of rice.

TABLE 2.4 GROUPING OF SAMPLE STATIONS ACCORDING TO
CLIMATIC REGION AND AVAILABLE WATER REGIME

Climatic Region	Available Water Regime Lasting Consecutively for				Total
	$\leqslant$ 3 months	4–6 months	7–9 months	$\geqslant$ 10 months	
I	1	2	5	9	17
II	0	1	1	10	12
III	0	0	0	8	8
IV	0	3	11	1	15
V	10	11	11	0	32
VI	0	0	6	0	6
VII	4	5	11	1	21
VIII	0	0	2	6	8
IX	6	0	0	0	6
Total	21	22	47	35	125

TABLE 2.5 GROUPING OF SAMPLE STATIONS ACCORDING TO
CLIMATIC REGION AND PONDED WATER REGIME

Climatic Region	Ponded Water Regime Lasting Consecutively for				Total
	$\leqslant$ 2 months	3 months	4–6 months	$\geqslant$ 7 months	
I	6	3	7	1	17
II	0	2	3	7	12
III	0	0	0	8	8
IV	0	0	12	3	15
V	25	4	3	0	32
VI	0	0	5	1	6
VII	12	4	5	0	21
VIII	1	1	4	2	8
IX	6	0	0	0	6
Total	50	14	39	22	125

Ponded water regimes lasting over six months would have some problems in drainage.

Using the results given in Tables 2.4 and 2.5, the following remarks may be made about each climatic region. Regions I, II, III, IV, VI, and VIII are generally suited for rice cultivation even under rain-fed conditions. Region I, the Strait-Sunda Region, is peculiar in that it has periods where surplus water is available for a considerable length of time but only a short period for ponded water, or none at all. This

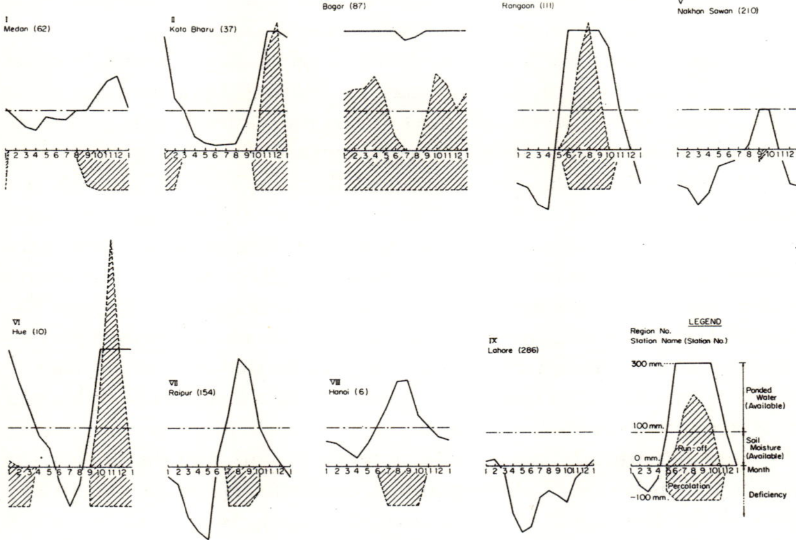

FIG. 2.4   PATTERNS OF SOIL WATER REGIME FOR DIFFERENT CLIMATIC REGIONS

type of water regime seems to be common in the southern half of the west coast of West Malaysia and is better suited to multiple cropping of upland crops. Furthermore, within Region I a subregion can be demarcated, separating the northern coast and the eastern half of the island of Java and the area farther east where a distinct dry season lasts for at least four months. But even here one crop of rice, depending solely on rainwater, is possible.

Region II, the Malay-Northern Borneo Region, seems to be suitable for one crop of rice during the rainy season and one upland crop for the rest of the year. In some areas cultivation of the second crop may be somewhat risky if no supplementary irrigation is provided.

Region III, the Oceanic Sumatra-West Java Region, has plenty of rain throughout the year and double cropping of rice should present no difficulties, even without a large-scale irrigation plan. In this region, upland crop cultivation may face drainage problems rather than irrigation problems.

The greater part of Region IV, the Southwest-Facing Coastal Region, experiences a distinct dry period lasting for more than four months. One crop of rice can be safely cultivated, but a second crop is not cultivable unless some form of irrigation is provided.

Region VI, the Middle Vietnam Region, is similar to Region IV in its soil water regime, but the dry period is not as severe.

Region VIII, the Tongking-Assam Region, has a favorable water regime for one crop of rice and may be followed by a planting of an upland crop. Here low temperatures during the winter months pose some difficulties for the cultivation of certain crops.

In contrast to the above regions, Regions V, VII, and IX are generally not suited to rice cultivation due to their unfavorable soil water regimes. Twenty-nine of 32 stations in Region V, the Southern Indochina-Southern India Region, 16 of 21 in Region VII, the Central India-Northern Indochina Region, and all 6 stations in Region IX, Lower Indus Region, are either prohibitive or marginal for rice cultivation under rain-fed conditions. Important rice areas occurring in these regions are either completely irrigated, as in Region IX, or located in water conservation areas, mostly at the mouths of great rivers where the land is naturally inundated for three to four months during the cropping season due to physiographic conditions. However, the greater part of the rice lands in Cambodia, Thailand, and India, countries which belong to Regions V and VII, are neither completely irrigated nor located within water conservation areas, and thus have a very unstable water supply. The considerable year-to-year fluctuation of the planted area for rice and/or of large areas of damage, can be cited as evidence of this instability caused by unfavorable soil water regimes.

The preceding computations and discussions are based on a long-term average of meteorological data. It is well known that the monsoon tropical climate is quite unstable, and the timing and the amount of rainfall can fluctuate greatly from year to year. Thus, the soil water regime as illustrated in Figure 2.4 should be regarded as one of the more frequent patterns probable at the site.

Though we have made use of Thornthwaite's method to assess the soil water regime, there is no guarantee of its validity when applied to a tropical climate, as Thornthwaite (1948) himself admitted. Kayane (1971) stated that the variation of evapotranspiration in the tropics is not temperature-dependent, but rather dependent on cloudiness, humidity, and other factors. There are some other methods devised for assessment of potential evapotranspiration, but the data needed for these latter methods are usually not available. Therefore, for the time being, Thornthwaite's method remains the only way to make approximations for the purpose.

The validity of the assumptions we have made must also be confirmed. The first assumption of 100 mm of water retention by the soil has often been questioned. The amount of water retention varies with texture, content of organic matter, and structure of the soil. In view of the wide variability in these soil properties, even within one climatic region, the assumption of 100 mm of water retention appears to be the best approximation.

Percolation loss is set at 100 mm per month, assuming a difficult-to-permeate subsoil layer with a hydraulic conductivity of 4×10^{-6} cm/sec. If permeability is better than this, the percolation loss is greater, and the water regime tends to be more unfavorable except in cases of excessive rainfall, such as in Region III. In most rain-fed rice lands the percolation loss is generally greater than assumed.

Solar Radiation

In reference to the climatic regions, we will consider another climatic factor, solar radiation, that affects rice cultivation, though not to the same extent as do water conditions.

Fukui (1973) prepared Figure 2.5 by superposing the solar radiation pattern on the rainfall pattern, taking the data from a study by Löf *et al.* (1966). In the figure, solar radiations for Regions III, VI, and VIII are not shown because sufficient data was not available. From the figure it is clear that in the lower latitude regions (I–V), solar radiation is solely determined by cloudiness, which parallels rainfall, ranging from some 400 cal/cm²/day in the rainy season to about 600 cal/cm²/day in the dry season. In the higher latitude regions (VII and IX), it is also affected by the season. Solar radiation goes down to 400 cal/cm²/day or even less during the winter months, although the weather is fine and dry.

There is a dilemma in rice culture; when water conditions are adequate, solar radiation is unfavorable. If water is made available artificially during the dry season, the dilemma is solved and a high yield may be expected. The highest yield of IR-8, the so-called miracle rice variety, at the International Rice Research Institute farm is around six and nine or ten tons of paddy per hectare in the wet and dry season, respectively. As the technological factors for both the seasons, as well as the temperature and day-length factors, are practically identical the difference in yield may be explained solely by the difference in solar radiation of about 100 cal/cm²/day between the dry and the wet seasons. The same explanation would apply to the remarkable success in the "green revolution" for rice in Pakistan, which is situated in Region IX, and Pakistan practices rice cultivation completely under irrigation.

2.2 PHYSIOGRAPHY OF TROPICAL ASIAN LOWLANDS

In the preceding section we considered the soil water regime from the climatological point of view. The actual water conditions in rice

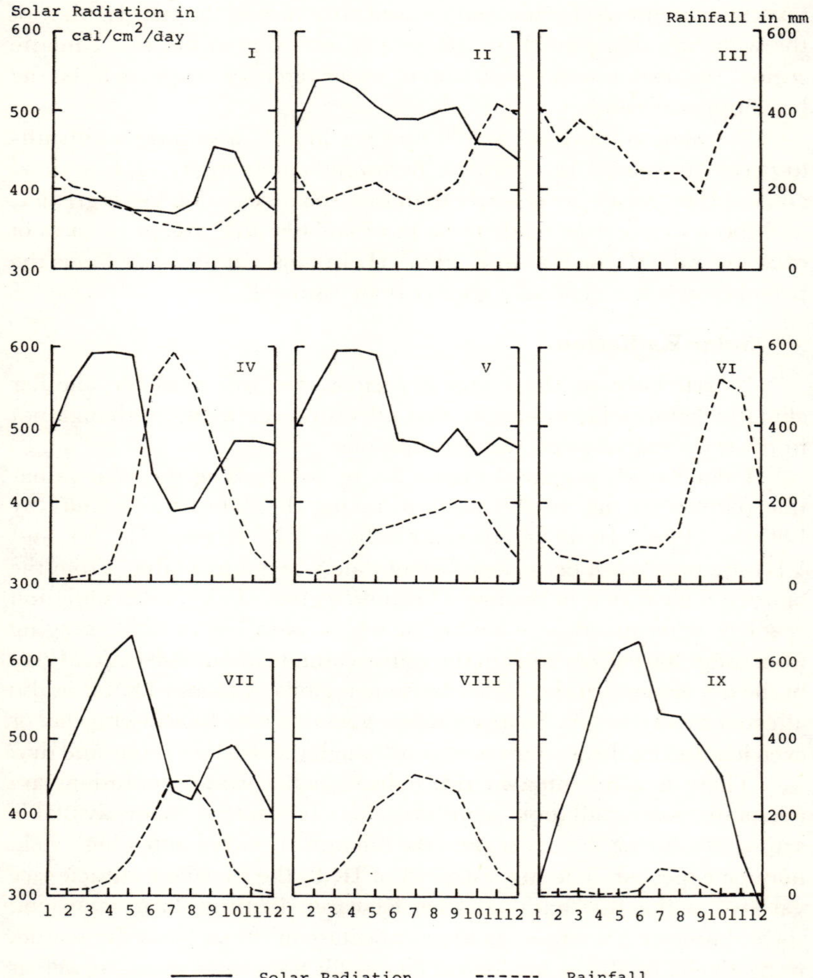

FIG. 2.5   MONTHLY RAINFALLS AND SOLAR RADIATIONS OF NINE CLIMATIC REGIONS IN SOUTH AND SOUTHEAST ASIA. (KYUMA 1972, and LÖF ET AL. 1966)

lands, however, are governed to an even greater extent by physiographic conditions.

First let us look at a map showing seasonally flooded areas. (Figure 2.6), which was drawn based on topo-maps and our field observations. The deltas of such great rivers as the Ganges-Brahmaputra, the

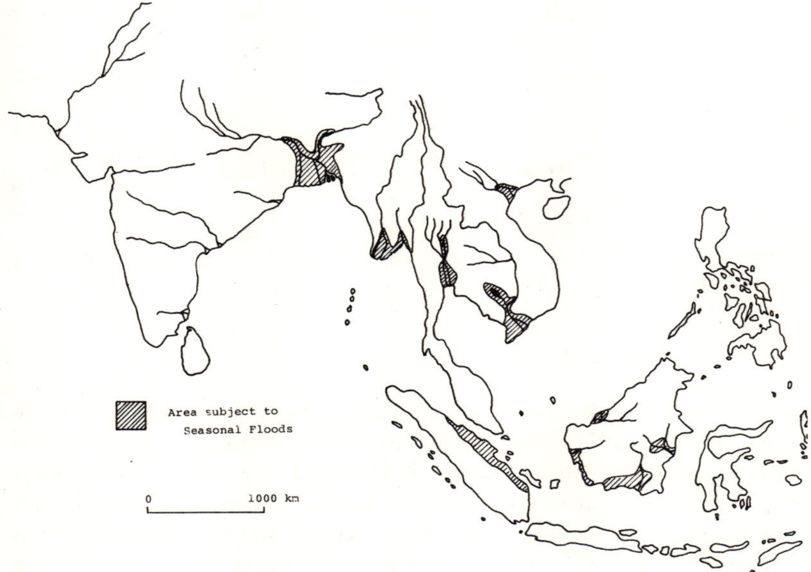

FIG. 2.6   MAP SHOWING SEASONALLY FLOODED AREAS.

Irrawaddi, the Salween, the Mae Nam Chao Phraya, and the Mekong, are flood areas, together with the coastal lowlands of insular Southeast Asia, notably the east coast of Sumatra, and the south and west coasts of Kalimantan. These areas suffering yearly floods are by definition alluvial lands, and extensive alluvial lands is one of the physiographic features of tropical Asia.

Table 2.6 gives data to endorse this statement (White House, 1967). Here the area of alluvial soil and the total land area of the world, Asia, and tropical Asia are given. For the world as a whole, alluvial soils occupy less than one-twentieth of the total land area, while in tropical Asia the figure is one-sixth. If we take only potentially arable land, that in alluvial soil areas is one-tenth of the total for the whole world, while it is one-third of the total for tropical Asia. In other words, more than one-third of the world's potentially arable land in alluvial soil areas is located in tropical Asia, which occupies only one-sixteenth of the world's total land area.

From the foregoing one can conclude that utilization of alluvial lowland is necessary in tropical Asia, where one-third of the potentially arable land is located in such lowlands. Adoption of rice as a crop may be regarded as an adaptation to the physiographic environment in tropical Asia.

TABLE 2.6 IMPORTANCE OF ALLUVIAL LAND
(in million hectares)

	LAND AREA		ALLUVIAL SOIL AREA	
	Total	Potentially arable	Total	Potentially arable
World	13000	3152	588	316
Asia*	2704	620	–	192
Tropical Asia	987	344	168	114

*Excluding USSR
(Source: World Food Problem, White House, 1967)

We now look at the physiographic features of rice lands in more detail. Takaya (1971) has recently made a physiographic study of rice lands in the Chao Phraya basin of Thailand. He defined six physiographic regions, that is, intermontane basins, constricted river channel area, old delta, delta flat, deltaic high, and fan-terrace complex as seen in figures 2.7 and 2.8, and summarized the characteristics of these six physiographic regions in Table 2.7. Fukui (1971) superposed his rice cultural regions on Takaya's physiographic regional division and discussed the features of rice cultivation in each region. We will give a brief description of each of the physiographic regions set up in Thailand in relation to rice cultivation, after the studies by Takaya and Fukui.

The intermontane basins are situated in North Thailand. They are characterized by high elevation, steep general slopes, strong local relief, and small area. The complexity of landforms or the occurrence of different physiographic units, such as recent alluvial plains, recent fans, fan-colluvial complexes, low and high terraces, within an intermontane basin is in striking contrast to the other physiographic regions, which are more homogeneous. In this physiographic region, most paddy fields are found either on recent alluvial plains, low terraces or recent fans. The high catchment/paddy area ratio is favorable to irrigated paddy growing, and, moreover, the small size of the rivers and the moderate general slope make water management feasible at a communal level. Thus, rice cultivation in the intermontane basins is thought to have started in early times.

The constricted river channel area forms low depressional stretches along the Nan and Yom rivers with an average width of about 20 km. This area is characterized by annual prolonged deep flooding, which is caused by the bottle-necked gorge on the Chao Phraya near the confluence of the Nan and the Ping. The prolonged deep flood during

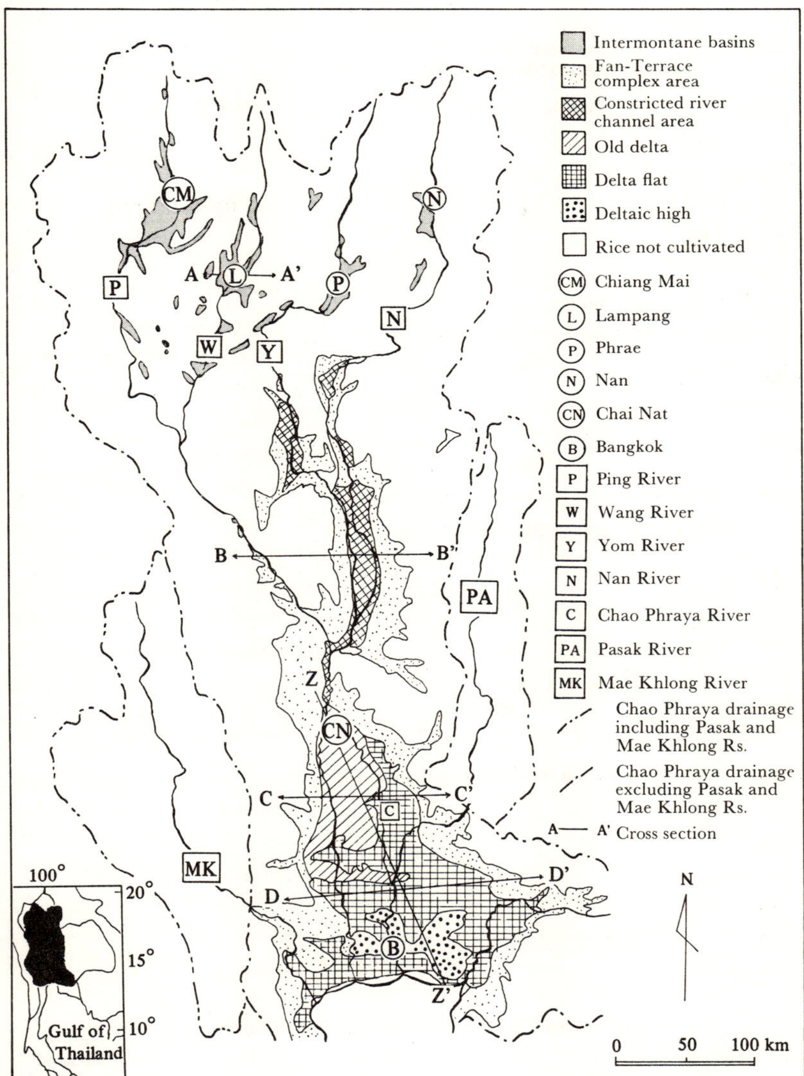

FIG. 2.7 PHYSIOGRAPHIC CLASSIFICATION OF PADDY LAND OF THE CHAO PHRAYA RIVER BASIN. The cross sections and a profile are shown in Fig. 2.8 (TAKAYA, 1971).

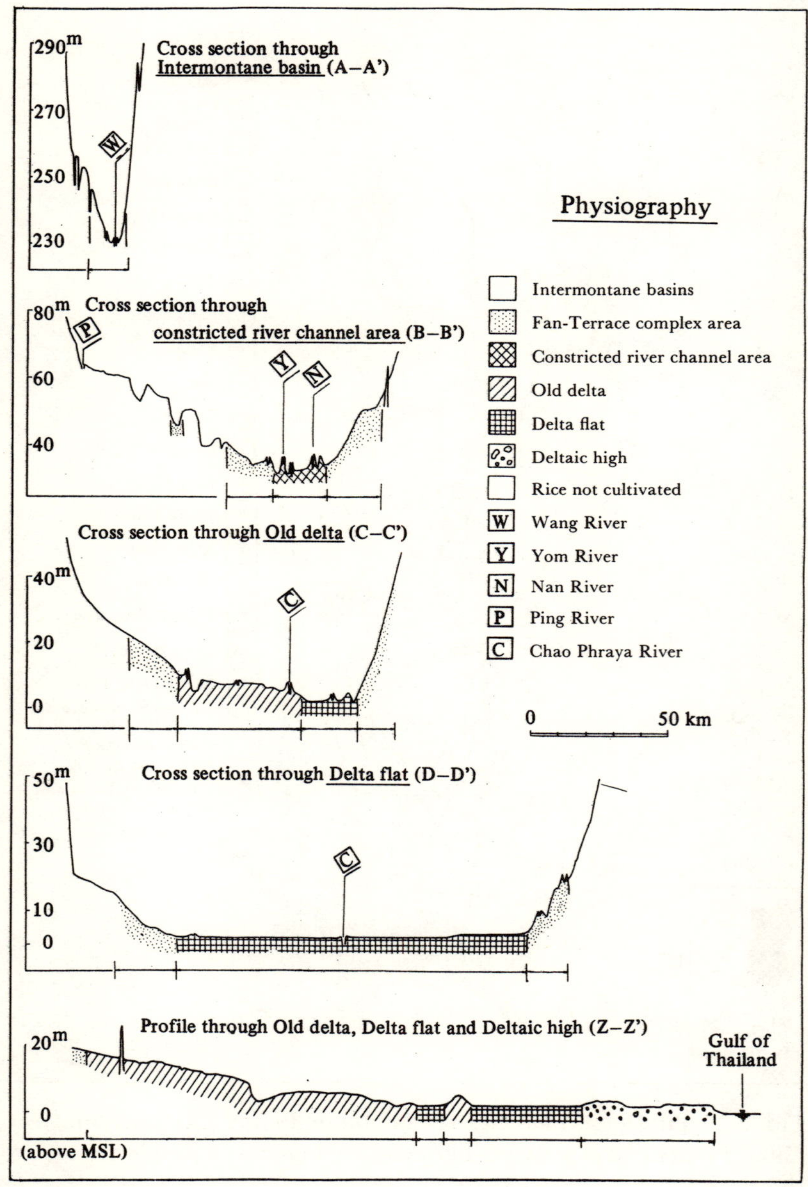

FIG. 2.8 CROSS SECTION AND PROFILE OF THE CHAO PHRAYA BASIN. The locations are given in Fig. 2.7 (TAKAYA, 1971).

TABLE 2.7   CHARACTERISTICS OF THE SIX PHYSIOGRAPHIC REGIONS

Physiographic Region	Elevation (m above M.S.L.)	General Slope (m/km)	Local Relief (m)	Acreage × 1,000 ha		Soil Texture	Geomorphic Setting	Catchment/ Paddy Area Ratio
				Gross	Net			
Intermontane basin	150–350	1.7 <	< 10	250	200	Clay to gravelly sand	Complex of stream alluvium, fan, and terraces	10 ∼ 40
Constricted river channel area	23– 60	0.2	± 5	320	200	Clay with a little sand	Recent alluvial plain	32.8
Old delta	5– 20	0.15	< 8	490	400	Clay with a little sand	Upper Pleistocene delta	23.0
Delta flat	0– 2	0.01	negligible	1110	820	Clay	Interdistributary low of Recent delta	15.6
Deltaic high	2– 4	0.01	negligible	210	160	Clay with a little sand	Relatively elevated parts in Recent delta	
Fan-terrace complex area	5–100	1.0–2.5	< 10	1800	1350	Sand with a little clay and gravel	Recent and Pleistocene fans and terraces	5.1

From Takaya (1971)

the rainy season sets a severe limitation on land use. Floating rice is the only cultivable variety in most areas of this region. Drainage rather than irrigation is the problem.

The old delta region occupies the upper part of the present-day Bangkok Plain. The Chao Phraya's distributaries begin somewhere near Chai Nat, and the deltaic nature of the plain is obvious. Geomorphologically, this is an upper Pleistocene delta partially superposed by recent levee deposits. The general slope is very gentle, but local relief is moderate, reflecting the normal dissection and the superposition of recent deposits. Because of the local relief, some parts of the region are too deeply flooded, and some parts are not submerged naturally. Today this region is under a government-administered large-scale irrigation scheme, and the greater part has become cultivable for rice, even for double-cropping if only sufficient water is available. Water conditions in the depressions, however, have not been improved, and only traditional varieties of floating rice can be cultivated.

The delta flat region is a recent delta formed by the Chao Phraya and has a very flat surface with negligible local relief. This region looks like a big lake during the rainy season, while it is a dry treeless plain during the dry season. Since the mid-nineteenth century many canals have been dug to make this area inhabitable, and, as a result, it has become the rice bowl of Thailand. From the point of view of soil, the brackish to marine depositional environment deserves special attention. Because of its brackish origin, the greater part of the area (about 800,000 hectares) has acid sulfate soils (see chapter 9).

The deltaic high is a slightly elevated region near the coast. Though the genesis of this region is not quite clear, the ground surface is about 1 m higher than the surrounding delta flat. Except for this difference in ground height, the landscape is similar to the delta flat. For rice cultivation, however, this slight difference is of great importance and more intensive farming practices, such as transplanting, are feasible. Soils are of marine deposit origin, and there are no problems with acid sulfate soils.

The fan-terrace complex region occurs along the margins of the Central Plain, where the plain proper merges into the mountainous region. Geomorphologically the area is composed of recent and upper Pleistocene fans and terraces. What is most characteristic of the region is the low catchment/paddy area ratio, which is estimated to be around five. Therefore, the area is basically water-deficient under the prevailing climatic conditions of Thailand. When the paddy fields are limited to small areas, however, the terrain is similar to the intermontane basins with respect to the ease of water management. Thus some of the fans have a long history of rice cultivation and have water distribution system that was constructed in earlier times. But as the paddy area is expanded, water deficiency becomes acute, and the greater part of rice lands in this region today is rain-fed. The conditions we described previously for the soil water regime of Region V apply to this fan-terrace complex region. According to Fukui's estimate (1971) the area of this physiographic region amounts to 1.3 million hectares or about 40 percent of the paddy lands in the Chao Phraya basin.

Most of the physiographic regions delineated in the Chao Phraya basin seem to occur as well in other big river basins. Even the constricted river channel area, the occurrence of which seems to be determined by a particular geologic structure in Central Thailand, has its counterpart in other regions, for example, in Northeast Thailand along the Mun and in Assam along the Brahmaputra. Only the deltaic high region is not encountered as an independent unit in other deltas. The delta flat and the deltaic high, therefore, should be treated as one region, the young delta region, in a general discussion.

Although the fan-terrace complex region occurs extensively throughout tropical Asia, its potentiality in rice production varies from one climatic region to another. In Regions V and VII water deficiency is acute, as in case of Thailand; and rice cultivation is inevitably unstable. But in Regions III and IV, for example, the fan-terrace complex region receives plenty of rainfall during the growing season, even rain-fed rice cultivation can be quite stable. If we wish to make our discussion comprehensive, we may have to add a few more physiographic regions. Takaya (1972) and Takaya and Tomosugi (1972b) have tentatively proposed two additional regions; one is a coast region, and the other is a plateau region. The former is a thin strip of young delta along the coast and falls into three categories: beach ridges, swales between beach ridges, and silted lagoons. Due to its low-lying topographic position, the region is inundated for a considerable period of the year, but not too deeply because water is freely drained into the nearby sea. Rice can be grown without sophisticated water control measures. Possible problems in rice cultivation in the coast region are the brackish nature of the deposits that give acid sulfate soils when reclaimed and direct salinization by sea water.

The plateau regions occur in Northeast Thailand, Upper Burma, and in India. They are dissected Plio-Pleistocene surfaces or uplifted peneplains and usually have undulating to rolling topography. Rice is grown in the depressions. As the catchment/paddy area ratio is even lower than that of the fan-terrace complex region, climatic control on the water condition of rice lands is more direct. The fact that these plateau regions are all in climatic Regions V and VII explains the very unstable rice cultivation in these regions.

2.3 RICE CULTIVATION IN TROPICAL ASIA

As stated earlier, rice cultivation in tropical Asia is thought to have developed as the adaptation of human beings to the natural environment, in which climate and physiography are of prime importance. According to recent archaeological evidence in Thailand, rice cultivation seems to have begun in North and Northeast Thailand sometime around 4000–3000 B.C., or even before that (see, Watabe, 1975).

The rice cultivated in those days, however, was not like the rice which we see today in the lowlands of tropical Asia. The archaeological finds are always at hilly sites, and the rice is thought to have been cultivated on the hillside slopes without flooding the fields. We know of similar rice cultivation practices by shifting cultivators today. The rice cultivated in this way is called upland rice, in contrast to lowland rice cultivated in flooded paddy fields. Thus, rice in early times was most probably similar to the upland rice of today. To be exact, however, it is better to say that the rice cultivated in early times was neither upland nor lowland rice by today's definition. Watabe (1975) supposes that early rice was adaptable to a wide variety of soil moisture conditions and would have grown in lowland (or flooded) conditions as well as on uplands. As experience grew, people started to recognize that lowland conditions gave more stable production and thus different varieties of rice were gradually differentiated for upland and lowland cultivation.

Using the lowland rice varieties, people started to inhabit such areas as intermontane basins and small fans, where water was easily manageable with their technology. Later as social regimes and technologies developed, the water of larger rivers could be controlled by men, and the plains and finally even the deltas were turned into rice lands. This last stage started only in the mid-nineteenth century.

In this assumed process of development of rice cultivation, the most crucial point was obviously the selection of lowland cultivation rather than upland cultivation. We believe that this choice was closely related to the climatic and physiographic conditions of tropical Asia.

First of all, we have to consider the pattern of rainfall in the region. The places which early inhabitants occupied were located in climatic Regions V and VII, where alternate dry and wet seasons prevail. The effects of climate on upland farming are twofold: (1) rain falls intensely for half the year, and erosion control becomes a serious problem; and (2) the weathering of upland soils is severe in the humid to subhumid tropical climate, and the maintenance of soil fertility is difficult.

During the rainy season, monthly rainfall amounts to 200 mm or more in many places, and this rain often falls in heavy squalls. An example is cited from Mohr and van Baren's (1954) *Tropical Soils*. In Indonesia, rains falling in "cloudbursts" amount to 8–37 percent, or an average of 22 percent, of the total annual rainfall in Indonesia, while in Bavaria (western Europe) comparable figures range from 0.5 to 3.7 percent, or an average of 1.5 percent. "Cloudbursts" are defined as rains having an intensity of > 1 mm/min. and lasting not less than 5 minutes. We must also give absolute annual rainfall. It ranges from 1200 to 3000 mm for Indonesia and less than 1000 mm for Bavaria. We can imagine how destructive the rains are in monsoon tropical Asia compared with those of western Europe. The difficulty of erosion control for people who started upland farming in hilly country is obvious. They would have been compelled to shift their site of cultivation fairly frequently.

The second factor enhances the need for such a move. Where topography is level and the risk of erosion small, the effect of weathering and leaching accumulates, and the soil fertility rapidly declines. As will be pointed out in chapter 3, even the upper Pleistocene terraces are strongly weathered and leached and show a certain degree of ferrallitization in the form of pisolitic iron concretions. In such a situation it would be difficult to continue upland farming, and long-term fallowing would be necessary to restore the soil fertility, which, in fact, means shifting cultivation.

Though upland rice has an advantage over other upland crops in its tolerance of wet injury, the danger of which is obvious in the relatively level part of monsoon Asia, it cannot tolerate erosion and fertility deterioration.

The adoption of lowland rice changes the picture drastically. For lowland rice cultivation it is necessary to make terraces when the terrain is sloping and to put levees around each field to flood it. These practices greatly diminish the erosion hazard even on slopes. The flooding naturally decreases the danger of drought, which is especially serious for upland crops during the first two to three months of the rainy season, when there are frequent dry spells. These points in themselves are quite important, but the greatest merit of the adoption of lowland rice cultivation is that it has made the utilization of annually inundated lowlands possible. As stated earlier, such lowlands are extensive in tropical Asia, and lowland rice has turned this vast area into productive arable land.

The geomorphic process prevalent in the alluvial lands, where most paddy fields are located, is depositional, in contrast with erosion in

TABLE 2.8   COMPARISON OF AMOUNTS OF BIOLOGICALLY ASSIMILATED NITROGEN IN DIFFERENT ECOSYSTEMS

ECOSYSTEM	FIXED NITROGEN in kg/ha
Upland field	7–28
Paddy field	13–99
Grassland	
Nonleguminous	7–114
Legumes + Grass	73–865
Forest	58–594

From Hauck (1971)

the uplands. Silt and clay, together with adsorbed bases, are carried down from the uplands to the lowlands and deposited there. Soluble mineral elements leached out of upland soils therefore are supplied to the plants growing in lowland soils. Thus the mineral nutrient status of lowland soils is moderate to high and remains semipermanently undepleted. A marine and brackish depositional environment is particularly favorable in this respect, except when a high amount of oxidizable sulfur (the causal material of acid sulfate soils) is present. In addition, there is favorable microbial action assisting the natural processes. As will be stated later in more detail (see chapter 3), in waterlogged soils two autotrophic nitrogen fixers are active, blue green algae and photosynthetic bacteria. In addition, some recent studies reveal the importance of hetrotrophic rhyzosphere bacteria in nitrogen fixation. We can see the effect of these microbes from the Table 2.8, which gives rough figures of the amount of natural nitrogen supply in different conditions. Clearly paddy field conditions are more advantageous than ordinary upland conditions. Ready release of phosphorus in the reduced condition of paddy fields is another advantage of lowland rice cultivation (see chapter 3).

In short, lowland rice cultivation has two definite advantages over upland crop cultivation;

ease of soil conservation, and

ease of maintenance of soil fertility.

These advantages were more important in the early days of human agricultural activity, when the level of technology was low.

Here we cite some concrete evidence showing the advantages of lowland rice cultivation. In northern Italy, rice was first cultivated in the fifteenth century. There is a record from that time which tells us that

TABLE 2.9   YIELDS OF RICE AND WHEAT IN DIFFERENT
CONTINENTS AND COUNTRIES

AREA	RICE (kg/ha)		WHEAT (kg/ha)	
	1961–1965	1972–1974	1961–1965	1972–1974
Africa	1722	1702	834	1007
North/Central America	3192	3803	1592	1950
South America	1731	1807	1361	1319
Asia	2049	2354	902	1211
Europe	4661	4578	2077	3124
Oceania	4568	5579	1245	1184
USSR	2461	3853	964	1537
Developed countries	4913	5616	1737	2175
Developing countries	1626	1867	973	1198

(Source: FAO Production Yearbook, 1974)

the harvested grains to seed ratio for wheat was only five, while the same ratio for rice was twice as high (Ciferri, undated). This means that at a comparable level of technology, rice is much more productive than wheat. The same is true even today. Table 2.9 is from the FAO Production Yearbook (1974), which indicates yields of rice and wheat on different continents and in different years. After allowing for milling losses of 22 percent for wheat and 35 percent for paddy, the yield of rice is, without exception, higher than that of wheat. The difference in productivity of rice and wheat on a unit of arable land is even greater in favor of rice, if we take the feasibility of continuous cropping of these two crops into consideration. Rice has been cultivated continuously for hundreds of years in the same field in many places in tropical Asia, and yet the yield has been kept at the moderate, if not very high, level of some 1.5 tons/ha. On the contrary, continuous cropping of wheat is possible only when a low yield level is accepted. Fallowing or crop rotation is common practice in many wheat-growing areas. In the long run, therefore, production of rice from unit land area is easily two or three times that of wheat. This high productivity of rice is certainly related to the high population density of tropical Asia, as either cause or effect.

3

General Features of Paddy Soils in Tropical Asia

The term "paddy soil" is not a pedological term, but signifies a type of land use. As the word "paddy" comes from "padi," which in Malay means rice, it is synonymous with "rice soil." In this chapter we will discuss general pedological and edaphological features of paddy soils in tropical Asia.

3.1 PEDOLOGICAL FEATURES

Occurrence of Paddy Soils

Paddy soil can occur on any type of soil in pedological terms. But as its occurrence is strictly conditioned by the availability of water, the types of soil most frequently found in rice land are naturally limited. By and large, paddy soils occur in low-lying lands which are inundated naturally or where water can be introduced by gravity. Therefore, most paddy soils occur on alluvia of the most recent geological periods, that is Holocene and the uppermost Pleistocene.

Thus most paddy soils are either Entisols and Inceptisols in the United States system of soil classification, or alluvial soils and such immature soils as humic gley and low humic gley soils, in the ordinary terminology.

Besides these, many other soil groups are cultivated for rice, although to a much lesser extent. Dudal (1958) once listed the following soil groups as occurring in rice lands: alluvial soils, gray hydromorphic soils, grumusols, latosols, andosols, regosols, red-yellow podozolic soils, planosols, and gray brown podzolic soils. Red and brown mediterranean soils or reddish-brown earths, and gray podzolic soils (see Dudal and Moormann 1964), are also utilized for paddy growing. Recently extensive areas of organic soils (peat soils) along the coasts of the insular part of Southeast Asia are being reclaimed for rice cultivation. If one includes soils under irrigation, then even soils in arid climatic conditions can be added to the list. Therefore, temperature is the final strictest constraint in land utilization for rice cultivation, but this is not a limiting factor in the tropics except in a high altitude area.

Thus, almost all kinds of soils can be used for rice cultivation if the water conditions are right. The potential of one soil group for rice cultivation, therefore, varies with different climatic and topographic conditions. For example, latosols in West Java in climatic Region III can be readily cultivated, but a latosol in Central India or in Central Thailand, in climatic Regions V and VII, cannot usually be utilized for paddy growing.

Specific Morphology of Some Paddy Soils

Soils that have long been cultivated for rice have been known to sometimes acquire special morphological features. For the cultivation of rice, soils are kept submerged, either naturally or artificially, for at least three to four months a year, and some management practices are conducted under these waterlogged conditions. Such treatment brings about special chemical and physical changes in soils under rice cultivation.

Reduction of the surface plow layer is the first change to be mentioned. Soils cultivated for rice have relatively high amounts of decomposable organic matter in the form of roots, stubble, and straw. When such a soil is waterlogged, the raw organic debris decomposes, and this is accompanied by microbial consumption of the dissolved oxygen. Thus, the soil becomes anaerobic and the ferric and manganic ions are reduced to ferrous and manganous ions of higher mobility. (For details of the chemical processes in submerged soils, see Ponnamperuma, 1972.) As will be discussed later in more detail, this process of reduction bestows upon the rice plants essential nutrients; ammoniacal nitrogen is liberated from decomposing organic matter and more phosphorus is available as a result of the reduction of ferric compounds.

Where there is moderate downward movement of water through the solum, the ferrous and manganous ions thus mobilized are leached from the surface plow layer into the more oxidative environment of the subsurface and subsoils and are then precipitated to form illuvial horizons. As the process is governed by their respective redox potentials, manganic and ferric oxides tend to precipitate separately, forming an iron accumulation zone which is underlain by a manganese accumulation horizon. A typical morphology of this type has been reported in Japan and elsewhere, notably in West Java, Indonesia, in "*sawah* soils" (Koenigs, 1950). In many cases, however, manganese is so readily mobilized even in upland conditions that the parent material for paddy soil formation is poor in manganese, especially in the tropics, and the typical horizon sequence of Apg/Birg/Bmng/Cg (or G), occurs rather rarely.

Plowing of this type of soil under waterlogged conditions often forms the so-called plow-pan or plow-sole right below the surface plow layer. This part of the solum undergoes compaction and usually preserves its gleyed color long after the draining of surface water. In a well developed plow-sole a thin platy structure is observed. This platy structure sometimes extends from the subsurface layer to the upper part of the illuvial horizon.

From the preceding, morphological features or the genetic horizons specific to the soils cultivated for lowland rice would be A-horizons eluviated under reductive conditions and B-horizons in which illuviation of iron (and manganese) has occurred under oxidative conditions. As the eluvial A-horizon is a prerequistite for the formation of the illuvial B-horizon, the presence of the B-horizon of oxidative illuviation serves as a necessary and sufficient criterion for detecting the specific morphology, which is acquired during rice cultivation.

From this it is evident that soils which have no illuvial iron accumulation just below the eluvial horizon do not, no matter how long they have been cultivated for lowland rice, form a separate soil group at any level of classification, distinct from their original soil groups.

Here arises the necessity to distinguish those soils having this specific morphology from paddy soils or rice growing soils in general. This point has not been clearly recognized in Japan until recently (Kyuma and Kawaguchi, 1966), and the term "paddy soil" has been used to signify both land use under rice cultivation, and a special soil group in general soil classification. Our proposed replacement for the latter usage of the "paddy soil" is "Aquorizem", and "Aquorization" for the process which leads to the specific morphology just stated.

In recent years the concept of aquorization has been developed by the studies conducted by Wada and Matsumoto (1973) and Mitsuchi (1974, 1975). Mitsuchi, making a comparative study of a paddy soil and an adjacent upland soil, elucidated that the aquorizem profile features are modified by permeability of the profile, which is in turn regulated by texture and pore size distribution. He distinguished three members of a "permeability series" of paddy soils or aquorizems: "brown lowland paddy soils", "gray lowland paddy soils" and "hanging-water gley lowland paddy soils", in order of decreasing permeability. The process that leads to the first group of soils is almost identical to that we explained in relation to a typical aquorizem. The profile patterns are as shown in Figure 3.1,A. In the second group, although iron-manganese illuvial horizons occur, the entire subsoil becomes gray-colored as a result of partial reduction of subsoil due to its slow permeability. Part of the iron and manganese is lost even from the subsoil, but the supply from the surface soil exceeds the loss, resulting in the formation of the illuvial horizons as a net result of the two opposite processes operating in the subsoil (see Fig. 3.1,B). The third group has a hanging-water gley horizon in the upper part of the profile due to an extremely low permeability, caused, for example, by a high expanding clay content. Weak eluviation of iron and manganese occurs from the hanging-water gley horizon and a weakly developed illuvial horizon is usually formed in the lower part of the profile (see Fig. 3.1,C).

It should be noted that all the soils in Mitsuchi's permeability series occur on relatively well-drained alluvial soils, such as Kamoshita's (1937) brown lowland soils and part of the gray lowland soils (see Table 3.1).

Wada and Matsumoto (1973) proposed a mechanism of subsoil "gray-coloring," as observed in many paddy soils. They postulate a downward movement of organic substances that have reducing ability and their interaction with the soil materials along their path through the profile, forming gray-colored ped and channel cutans. They further argue the role of the reducing organic substances in the separation of manganese illuvial horizon from the iron illuvial horizon.

Classification of Paddy Soils

Since the 1930s, attempts have been made in Japan to classify paddy soils. Kamoshita (1937) was the forerunner, and he attempted to fit paddy soils into Stremme's scheme of classification of "Nasse Böden." Inspired by the ideas of Uchiyama (1949) and Kanno (1956) in the early postwar period, Yamazaki (1960), Kanno (1962), Oyama (1962), the pedologist group (1969), and Matsuzaka (1969) have made several proposals since 1960. Correlation of the soil units of these authors at the great (soil) group level is shown in Table 3.1. Except for Oyama's and Matsuzaka's schemes, in which great groups are set up on consideration of such diagnostic horizons as gleyed horizon, gray-colored horizon, volcanic ash horizon, and so forth, all the others take hydrological conditions as the most important criterion, that is, the presence or absence and the intensity of gleyzation due to ground water, reduction of the surface horizon due to irrigation water, and the degree of migration of soil materials due to the percolation of surface water. In Fig. 3.2 catenary sequence showing paddy soil types in relation to ground water level is cited from Kanno (1956).

In reviewing these classification schemes, two related points can be questioned. One concerns the relationship of paddy soils to other soils in the general soil classification system, and the other the confusion in the usage of the term "paddy soil," which has already been mentioned. The first point can be illustrated by peat and muck paddy soils. These soils never show the morphological characteristics of aquorizem, yet they are included in the classification of paddy soils. So what is the relation between peat and muck paddy soils and peat and muck soils (organic soils) in the general classification? The same point can be raised in connection with the "ground water gley rice soils" of Kanno, with a horizon sequence of gleyed plow layer (Apg)/gley horizon (G). They are indistinguishable, morphologically, from ordinary gley soils. In short, it is difficult to classify paddy soils without incorporating them into the general scheme of soil classification. We have to elucidate the

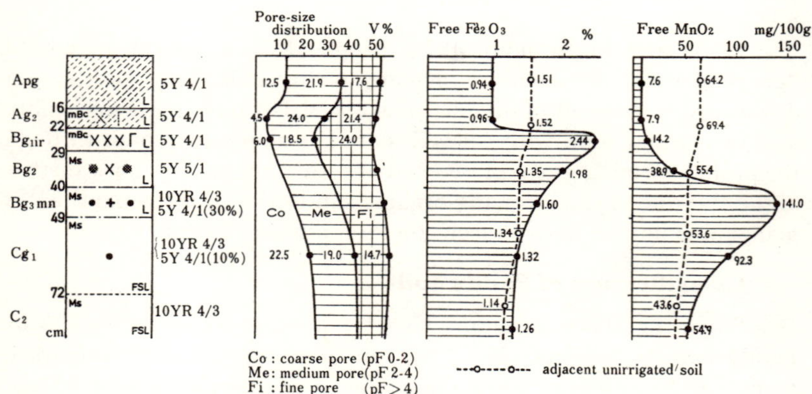

A. BROWN LOWLAND PADDY SOILS (SEKIJO, IBARAKI PREFECTURE)

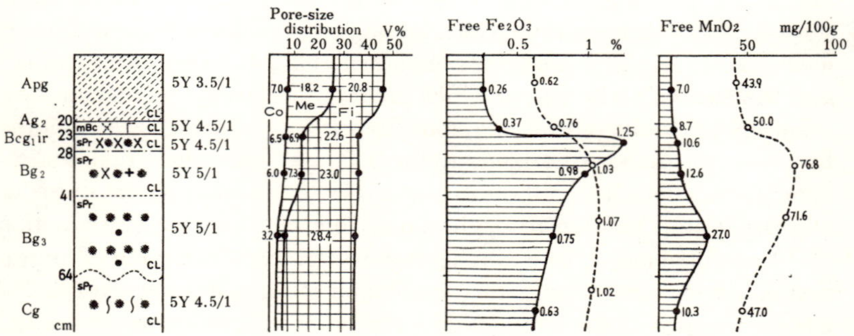

B. GRAY LOWLAND PADDY SOILS (TADOTSU, KAGAWA PREFECTURE)

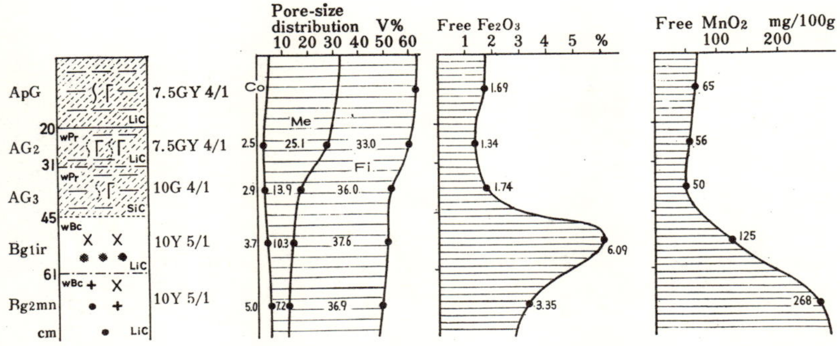

C. HANGING-WATER GLEY LOWLAND PADDY SOILS (AZAI, SHIGA PREFECTURE)

Key to A, B, C:
Rusty mottles
    shape . . . .   **X** threadlike, ferruginous
                                                          ʃ tubelike, ferruginous
                                                          ✱ cloudlike, ferruginous
                                                          + threadlike, manganiferous
                                                          ● spotlike, manganiferous
    abundance . . . .   **X** few   **XXX** plenty
                                                 XX common   **XXXX** abundant
    contrast . . . .   **X,ʃ** strongly contrasted
                                                       X,ʃ weakly contrasted

Structures
    shape . . . .   Bc; blocky
                                                        Pr; prismatic or columnar
                                                        Ms; massive
    development . . . .   w; weak, m; medium, s; strong

Strong gleyzation . . . .   ▬ ▬ ▬

Horizon boundary . . . .   —— sharp, —·—·— clear,
                                             ------- gradual

Organic matter . . . .   ▭ < 2%, ▨ 2–5%

FIG. 3.1 ILLUSTRATION OF PROFILE FEATURES OF THREE PADDY SOILS IN A PERMEABILITY SERIES (MITSUCHI, 1975)

proper position in general soil classification for paddy soils, which have undergone different degree of specific soil formation, that is, aquorization, and which show different degrees of deviation from the original soils.

Dudal (1965), taking note of the occurrence of "inverted gley" in paddy soils, considers that what we call aquorizem can be adequately described at the subgroup level and proposes the adjective "anthraquic" to be added to the great group name from which the soil is derived.

KAMOSHITA	UCHIYAMA	OYAMA	MATSUZAKA	KANNO	YAMAZAKI	PEDOLOGIST
Peat soil	Peaty freshwater paddy soil	Peat soil	Peat soil		Ground water paddy soil	Peat soil-paddy soil
Muck soil	Mucky freshwater paddy soil	Muck soil	Muck soil			Muck soil-paddy soil
Wet lowland soil	Freshwater paddy soil Blue reductive type	Gley soil	Strong gley soil Gray-colored gley soil Upland-gley soil	Ground water gley rice soil AI - - - - - AII - - - - -	(Gr.gr) (Gi.gr) (Gi.gi)	Gley soil-paddy soil Stagno-water gley-like semi-terrest. paddy soil Stagno-water gley-like terrestrial paddy soil (in part)
Gray lowland soil	Freshwater paddy soil Gray leached type Gray-brown Intermediate type	Gray-colored soil	Gray lowland soil Gray-brown lowland soil Wet gray-brown upland soil	Intermediate gley-like rice soil BIII - - - - - BIV - - - - -	(Ai.gi) Irrigation water paddy soil (Ai.a)	Stagno-water gley-like terrestrial paddy soil (in part) Stagno-water gley-like terrestrial paddy soil (in part) - - - - -

					(Al.a)	Surface-water gley-like semi-terrest. paddy soil
Brown lowland soil	Freshwater paddy soil	Yellowish-brown Colored soil	Yellowish-brown lowland soil Wet yel.-br. upland soil	Surface water gley-like rice soil	(Ai.b)	Surface-water gley-like terrestrial paddy soil
------------------	Brown oxidative type		------------------------	CV	(Al.b)	
Red lowland soil			Reddish-brown lowland soil Wet red.-br. upland soil			
	(Dark-colored soil type)	Black volcanic ash soil	Kuroboku low-land soil Kuroboku gley soil Wet kuroboku upland soil			Classified at lower categories
	------------------	Light-colored volcanic ash soil	Wet light-colored kuroboku upland soil			

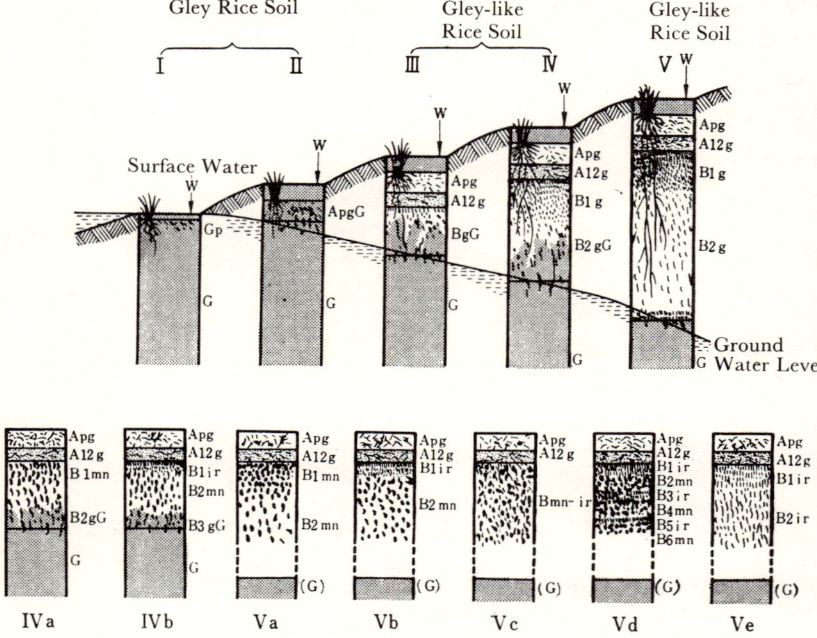

FIG. 3.2   SCHEMATIC CATENARY RELATION BETWEEN DIFFERENT PADDY SOIL TYPES (KANNO, 1956)

Karmanov (1966) considers, in his study on Burmese paddy soils, that when a qualitatively new process occurs in a soil through cultivation of rice, the soil is an independent genetic soil type. When a "red-brown soil" of the dry savanna is brought under rice cultivation, essential changes occur in the water regime, physical properties, and so forth, and such qualitatively new processes as degradation come into existence. (The term "degradation" used by Russian scholars is not clearly defined, but seems to mean the changes leading to a coarser texture and a bleached whitish appearance [Rozanov and Rozanova, 1965].) In this case a new genetic soil type, "red-brown meadow soil", can be distinguished. Likewise, in Burma, "reddish-yellow meadow soils" and "reddish-cinnamon meadow soils" are the genetic soil types of paddy fields. However, when a "dark meadow soil" is cultivated for rice, changes in the water regime, physical properties, and so forth, are not marked and are only quantitative. In such a case he considers that these soils under rice cultivation are not a new genetic type and should be separated on lower taxonomic levels from the original soils.

According to Fridland (1964) who studied paddy soils in North Vietnam, it is necessary to distinguish many types of paddy soils, each of which may be related either to the specificity of the original soil type or to the processes taking place in the soil under rice cultivation, for example, degradation. In North Vietnam he distinguishes the following genetic soil types or groups of genetic soil types: (1) ferrallitic rice soils, (2) alluvial-littoral rice soils, and (3) marshy rice soils.

Otowa (1969) recently discussed the morphological changes of soils under paddy rice cultivation, taking examples from Japanese paddy soils. In his conclusion he says that in order to accommodate some paddy soils with iron and manganese illuvial horizons, which are equivalent to aquorizem, it may be necessary to prepare "hydragric" subgroups in such original great groups as Fluvisols, Gleysols, and Cambisols. The same suggestion was made by Wada (1966). "Hydragric" is a modification of the term "agric" used in the U.S. system for paddy soil conditions.

Whatever name is chosen, a category equivalent to the subgroup in the U.S. system would be the proper place for paddy soils with a certain degree of specific morphological deviations. In view of the facts that the aquorizem morphology is a mere superimposition on the more fundamental soil nature and that the land use may be changed after some years, it seems inappropriate to set up an independent class above the subgroup level. But before finally settling this problem, more careful examination and characterization of the differentiating characteristics of aquorizems is necessary.

Morphological Characteristics of Paddy Soils in Tropical Asia

The greater part of the soils we investigated in tropical Asian countries do not have clear iron and manganese illuvial horizons and thus do not qualify as aquorizems. They remain as alluvial soils, low humic gley soils, and grumusols even after hundreds of years of rice cultivation.

At first it was difficult for us to comprehend the rare occurrence of aquorizems in areas where rice cultivation has a long history and where the fields appeared to be *kanden* (in Japanese, dry paddy fields) at the time of our field study, mostly during the dry season. *Kanden* in Japan are waterlogged artificially during the rice-growing season only and kept drained during the off-season for cultivation of a second crop or for fallow. This type of soil acquires the aquorizemic morphology in a very short period of time under Japanese conditions; for example, some moderately fine textured soils in the polder land of Kojima Bay formed B-horizons during forty to fifty years of rice cultivation, and in the case of irrigated upland soils the period can be even less than ten years.

In considering the paddy soils in tropical Asia, however, it should be borne in mind that most paddy fields are situated in the lowlands where the soils are naturally inundated with rainwater or flooded by river water. Thus, the topography does not favor the downward movement of water through the solum during the rice-growing season. This is especially true for most deltaic soils. The land is low-lying and the heavy rains accumulate on the plain around the big rivers making the ground water level as high as the ground surface. Both external and internal drainage is almost totally impossible.

Because of the very heavy texture of these deltaic soils many wide cracks occur during the dry season. The first few rains at the beginning of the rainy season may percolate fairly rapidly, but once the clay expands there is little or no possibility of further rains penetrating the solum. This contraction-expansion pattern of the clays is evidenced by slickensides and pressure faces which occur in the subsoil of many of the deltaic soils. As stated in chapter 5, except for grumusol and some grumusolic alluvial soil, the clay mineral composition of these soils is not necessarily dominated by montmorillonite. We now believe, therefore, that the presence of slickensides does not necessarily indicate a dominance of montmorillonite or swelling clays in the clay mineral composition.

Paddy soils on alluvial terraces are usually classified as low humic gley soils that underwent intense weathering before the present land use began. They usually have more or less lighter-textured surface soils and clayey, very compact, strongly mottled subsoils. With bunding for rice cultivation, the land surface is inundated during the rainy season, but dries up rapidly when the rains cease. The moisture condition in the subsoil, however, remains almost constant throughout the year and not much oxygen diffuses into the subsoil horizons even during the dry season because of the compactness of the soil. In such conditions, downward movement of water during the waterlogged period is strongly restricted, and even when slight percolation occurs there is little opportunity for ferrous and manganous ions to be deposited and form a conspicuous illuvial horizon with sharp boundaries. In addition, the general paucity of these soils in organic and mineral nutrients would restrict the development of reduced conditions in the surface layer because of poor microbial activity. The low iron and manganese content of the surface soil may also limit the formation of the illuvial horizons.

Some soils with an aquorizem morphology do occur under certain conditions in tropical Asia. They are without exception situated on terrain with good external drainage and their sola also have good permeability owing to either a lighter texture or a developed structure. Profile descriptions of three soils of this category are given below.

Soil M-36
Location: Kampong Belukar, Bukit Badak, 2 miles south of Bachok, Kelantan.
Landform and parent material: Upper part of "bris sand" or subrecent coastal sand ridge
Soil classification: Rudua Series
Apg   0–10 cm. grayish brown (2.5Y 5/2, moist) loamy sand; common yellowish red (5YR 4/8) and yellowish brown (10YR 5/6) tubular and filmy mottles; weak blocky; abrupt wavy boundary to
A12g   10–17 cm. grayish brown (2.5Y 5/2, moist) fine sandy loam; few yellowish brown (10YR 5/6) tubular mottles; single grained; somewhat compact; abrupt wavy boundary to
A2g   17–22 cm. light gray (2.5Y 7/2, moist) fine sandy loam; few yellowish brown (10YR 5/6) tubular mottles; single grained; somewhat compact; abrupt wavy boundary to
Blg   22–40 cm. mottled brownish yellow (10YR 6/8) and yellow (2.5Y 8/6, moist) fine sandy loam; single grained; somewhat compact; gradual smooth boundary to
B2g   40–60 cm. mottled brownish yellow (10YR 6/8) and yellow (2.5Y 8/6, moist) fine sandy loam with some dark concretionary spots; single grained; somewhat compact; gradual smooth boundary to
BCg   60–100 + cm. light yellowish brown (2.5Y 6/4, moist) fine sandy loam; many light olive brown (10YR 5/6) cloudy mottles; single grained.
Remarks: Ground water came up to 100 cm. Fine mica flakes increasing downward; below 110 cm greenish colored sand with many biotite and muscovite flakes.

Soil Bd-20
Location: Godrakura, P.S. Haluaghat, Mymensingh District; 2 miles north of Haluaghat, 300 m east of the road.
Landform and parent material: Middle part of gently sloping land on subrecent piedmont plain sediments
Soil classification: Jhinaigati Series
Apg   0–12 cm. gray (10Y 6/1, moist to dry) loamy fine sand with few white sand separations; common light

yellowish brown and brownish yellow (10YR 6/4 and 6/6) fine tubular, filmy and diffused tubular mottles; structureless; common large voids; abrupt smooth boundary to

Bg　　　12–26 cm. mottled gray (10Y 5.5/1, moist to dry) and yellowish red (5YR 4/8 and 5/8) sandy loam with common white sand separations; very weak, very coarse, blocky; many fine and small pores with mainly sand (very few clay) coatings; gradual smooth boundary to

BClg　　26–40 cm. gray (10Y 6/1, moist to dry) sandy loam with few white sand separations; many yellowish red (5YR 5/8) small (less diffused) cloudy mottles; weak medium to coarse blocky with clay coatings on ped surfaces; many fine and small pores with clay coatings; clear smooth boundary to

IIBC2g　40–57 cm. gray (10Y 6/1, moist) clay loam; many dark brown (7.5YR 4/4) (Mn?) and reddish yellow (7.5YR 6/8) small (less diffused) cloudy mottles; weak medium to coarse blocky with clay coatings on the ped surfaces; many fine and small pores with clay coating; gradual smooth boundary to

IIIClg　57–80 cm. gray (10Y 6/1, moist) sandy loam; many reddish yellow (7.5YR 6/8) small cloudy mottles with common dark brown (7.5YR 4/4) (Mn?) small cloudy mottles in the upper part; very weak coarse and very coarse blocky with few crack surfaces with clay and sand coatings; many fine and small pores with clay coatings; gradual smooth boundary to

IIIC2g　80–100 + cm. gray (10Y 6/1, moist) sandy loam; many reddish yellow (7.5YR 6/8) small and medium cloudy mottles; very weak very coarse blocky with few crack surfaces with clay and sand coatings; many fine and small pores with clay coatings;

Remarks:　Many fine quartz grains and mica flakes in the fifth and sixth horizons; minerals difficult to identify in the other horizons, but mica probably present.

Soil No. IN-1
Location:　Kedung Halang, Kedung Halang, Bogor, West Java
Landform and parent material:　Upper part of sloping land (in rolling country) on old volcanic fan
Soil classification:　Reddish-brown Latosol

Apg　　　0–15 cm. brown (10YR 4/4, moist) heavy clay with few medium rounded gravels; common but unevenly distributed gley spots (5Y 4/2) and common 7.5YR 4/6 filmy and fine tubular mottles; weak coarse blocky; friable; clear smooth boundary to

ABg　　　15–27 cm. brown (7.5YR 4/3, moist) heavy clay; many but unevenly distributed gley spots (5Y 3/3), few

	7.5YR 5/6 fine tubular and spotty mottles; moderate to strong medium and coarse subangular blocky with patchy Mn-clay coatings; slightly firm; abrupt smooth boundary to
B1g	27–46 cm. reddish brown (6.25YR 4/6, moist) heavy clay; many Mn fine tubular, filmy, and fine cloudy mottles; strong medium and coarse subangular blocky with Fe and Mn-clay coatings; firm and somewhat compact; gradual smooth boundary to
B2g	46–75 cm. dark brown (7.5YR 3.5/4, moist) heavy clay; many 7.5YR 6/6 veinlike mottles and common Mn filmy and fine tubular mottles; moderate medium subangular blocky with patchy Mn-clay coatings; friable; gradual smooth boundary to
BCg	75–100 + cm. dark brown (7.5YR 3.5/4, moist) heavy clay; common pale yellow veinlike mottles and few Mn thin filmy mottles; weak medium subangular blocky with thin patchy Mn-clay coatings; very friable
Remarks:	Dark colored ferromagnesian minerals are visible throughout the profile.

The first profile, on "bris sand" on the east coast of West Malaysia, very strikingly resembles some of the Japanese degraded *akiochi* paddy soils. The upper two horizons are dark humiferous Al horizons with iron oxide content of less than 0.5 percent. The third horizon contains about the same amount of free iron oxides but has a lighter color due to a paucity of organic matter. Because of this it looks like an A2 or bleached horizon of a typical podozol soil. Free iron oxide content abruptly increases in the fourth horizon and even more in lower horizons. Therefore, no bulge in the iron oxide content is seen within a depth of 1 m. Normally, illuvial iron accumulation in paddy soils does not occur below 50 cm from the surface, but in some Japanese *akiochi* paddy soils a similar gradual increase in iron oxide content to a greater depth is known to occur. This can be explained by the paucity of iron oxides in the parent material and the high intensity of leaching in these sandy soils.

The second profile was observed on the piedmont plain of the Shillon hills in northern Bangladesh. From the textural differences and the occurrence of unweathered mica flakes, two lithologic discontinuities may be recognized. The presence of sand separations (see the following section on Morphological Features Associated with Weathering) in the upper three horizons suggests that the soil is well aged. Clay and fine sand coatings were observed throughout the profile on ped faces and pore walls. Iron illuviation in the second horizon is conspicuous in the field but not in the laboratory because of the presence of many mottles in the third horizon.

The third profile was situated on an old volcanic fan in West Java.

Unlike the preceding two profiles, this is a heavy clay soil with reddish hue, a reddish-brown latosol in the Indonesian classification. But apparently the internal drainage is good owing to development of strong structure. The B1-horizon is the iron illuviation horizon and B2, the manganese illuviation horizon. In spite of the name "latosol," dark colored mineral grains are visible throughout the profile. This profile is similar to the one Koenigs (1950) described as *sawah* soil.

In both the second and third profiles ped faces and pore walls are modified by clay, sand, or iron and manganese. And this feature is not specific to aquorizemic profiles. Even in very heavy textured soils of deltas ped faces and pore walls of the subsoil often have clay coatings, some of which show clear signs of flow. Brammer (1970, 1971) also noted this in Bangladesh and called such coatings "flood coatings." He says that, "These coatings comprise both silt and clay, and they develop very rapidly. Their presence is not diagnostic for an argillic horizon. . . . The genesis of these flood coatings is not yet fully understood, nor is their significance in soil classification." We share Brammer's opinion.

Morphological Features Associated with Weathering

During our field surveys in tropical Asia we encountered many soils containing rounded iron concretions of various size and hardness which we call pisolitic concretions. In some cases they were the size of mustard seeds and relatively easily crushed between the nails, while in other cases they were very hard and of greater size. Sometimes even clustered pisolitic concretions (a kind of hardened laterite block) were present in the soil. Most of these pisolitic concretions were found in the subsoil together with prominent mottles. The genesis of these concretions was obviously connected with the age and degree of weathering of the parent material, because their occurrence was more frequent and their size and hardness more developed in soils on older terraces.

Some of the results of a recent study on these concretions are quoted in Table 3.2. (Furukawa *et al.*, 1976). Concentration of iron and manganese is quite variable; some are obviously richer in iron, while others are manganiferous. Iron minerals identified were, for the most part, goethite and rarely hematite. Manganese minerals were not identified. Silica came mostly from quartz, but some concretions contained a small amount of kaolin clay minerals and/or feldspars.

On closer investigation, however, the genesis of pisolitic concretions is seen to be not only a function of time but also a function of the chemical nature of the material, hydrographic conditions, and so forth. But in areas where the latter conditions are more or less similar, the time factor is the determinant of the size and hardness of these concretions.

TABLE 3.2   PROPERTIES OF PISOLITIC IRON CONCRETIONS OCCURRING IN PADDY SOILS OF TROPICAL ASIA

Soil No.	Lab. No.	Horizon	Color	Size (mm diam.)	Shape	Surface Texture	Luster	Hardness	SiO_2	Al_2O_3	Fe_2O_3	Mn_3O_4	CaO	MgO	TiO_2	K_2O	P_2O_5
T-13	(2)	4	7.5YR3/3	2–5	spherical/botryoidal granular	rough/ smooth	dull/ greasy	7	19.4	11.4	67.9	0.28	0	0.45	0.30	0	0.22
T-42	(7-1)	1–4	7.5YR3/3	5–8	botryoidal spherical	smooth	greasy	5	46.4	8.8	33.7	7.97	0.30	1.42	0.78	0.45	0.09
"	(7-2)	1–4	7.5YR3/3	1.5–4	spherical granular	smooth	greasy	5	28.6	7.5	61.4	0.83	0.08	0.46	0.40	0.28	0.38
I-36	(37-1)	2	10YR3/3	1.5–3	spherical granular	smooth	greasy	5	21.4	10.8	64.7	0.53	0.23	0.87	0.71	0.57	0.17
I-72	(40)	4	10YR5/8	2–7	spherical granular	rough	dull	3	62.3	9.0	26.7	0.22	0.33	1.03	0.66	0.01	0.01
Ph-27	(26-3)	1	10YR3/2	2–6	spherical granular	smooth	dull	5	23.9	6.4	64.1	3.14	0.06	1.76	0.49	0	0.05
Ph-39	(28)	1	10YR5/3	2–3	spherical/botryoidal granular	smooth	dull	3	39.8	12.9	37.9	5.14	0.43	1.57	0.88	1.06	0.19
In-2	(14-2)	6	7.5YR4/6	2–2.5	spherical/subspherical	rough	dull	2	24.1	15.8	52.9	4.40	0.01	0.63	1.38	0	0.70
In-34	(20)	3	7.5YR6/8	2–15	subangular blocky	rough	dull	2	22.4	22.3	52.4	0.56	0.14	0.93	0.89	0	0.28
Ph-27	(27-1)	2	10YR3/3	2–7	spherical granular	rough	dull	2	28.7	11.8	24.8	31.99	0.26	0.80	1.55	0	0.01
In-2	(14-1)	6	7.5YR4/6	5–8	spherical/subspherical granular	rough	dull	2	28.9	15.3	33.7	18.30	0.08	0.98	2.16	0	0.47

From Furukawa, *et al.* (1976)

Takaya (1968) regards the occurrence of pisolitic concretions as a key in identification of the stratigraphy of Quaternary sediments, on the assumption that these concretions are a product of subaerial weathering over a substantial length of time, and, accordingly, are an indication of unconformity. A schematic cross section of the Quaternary deposits of Thailand and descriptions of different physiographic units recognized by Takaya on the basis of soil features, among other factors, are given in Fig. 3.3.

Another specific feature encountered in tropical Asian paddy soils are what we call "sand separations." These sand separations are irregularly shaped sandy spots embedded in a more clayey subsurface horizon. These are typically found in paddy soils on older terrace surfaces, for example, the Madhupur Jungle Tract in Bangladesh. Similar formations were found in some paddy soils in Cambodia and India. Formation of these sand separations appears to be associated with the

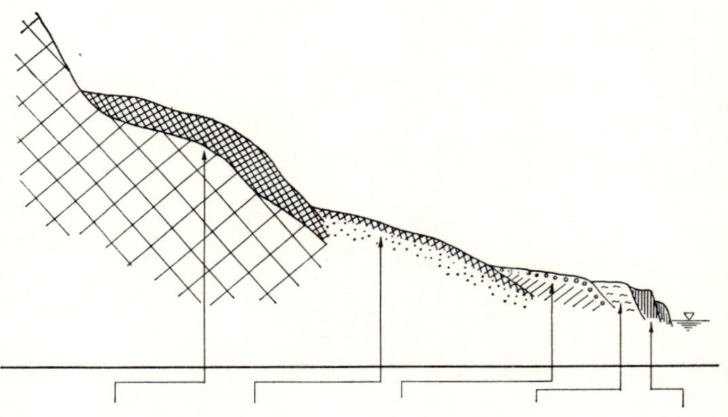

FORMATION	PENEPLAIN	TERRACE III	TERRACE II	TERRACE I	FLOOD PLAIN
SEDIMENT	No young alluvium	Dominantly sandy	Dominantly clayey	Loamy	Dominantly sandy
WEATHERING FEATURE	Thick hard laterite	Heavily weathered alluvium capped by thin laterite	Weathered alluvium with pisolitic concretions	Alluvium with concentration	Fresh alluvium with concentration
TOPOGRAPHY	Undulating to rolling	Slightly undulating to undulating	Flat to slightly undulating	Flat	Flat and slightly undulating
AGE	Lower Pleistocene	Middle Pleistocene	Upper Pleistocene	Holocene or Uppermost Pleistocene	Holocene

FIG. 3.3   SCHEMATIC CROSS SECTION OF QUATERNARY DEPOSITS IN THAILAND (TAKAYA, 1968)

length of time over which the soil material has undergone weathering and soil formation under rice cultivation.

Unfortunately, the process of formation of both the pisolitic concretions and sand separations is not yet known. The latter may be related to what Brinkman (1970) calls "ferrolysis," which is thought to be a soil-forming process specific to seasonally flooded conditions. On the genesis of pisolitic concretions many inferences have been proposed (confer, Kyuma, 1970), but no solid evidence has yet been obtained. In view of the importance of these concretions in the field judgment of the nature of a soil, the study of their genesis should be intensified.

EDAPHOLOGICAL FEATURES

Physical Properties of Paddy Soils

In comparison with soils cultivated for other crops, paddy soils have the following characteristics in relation to crop production; as the soils are waterlogged, either naturally or artificially for some months during the cropping season, the importance of the physical properties of the soil is relatively small as long as a sufficient amount of water is available. Therefore, the productivity of paddy soils is heavily dependent on soil fertility or the chemical nature of the soils, because the role played by soil tilth, which is bound to physical properties, is relatively unimportant.

The experience in Japan, however, tells us that a moderate rate of downward movement of water, say ten to twenty mm per day, is essential for a high yield of rice, for example, more than nine tons of paddy per hectare. This, first of all, removes from the rhyzosphere such toxic substances as butyric acid that are produced in reduction processes under waterlogged conditions. Furthermore, elaborate water management such as intermittent irrigation, which has been adopted by progressive farmers in Japan, is said to have good effects on rice plants and provide the rice roots with oxidative environments.

These examples suggest that physical properties do have some importance even in paddy soils. The yield levels aimed at in such cases, however, are much higher than the average yield actually attained today by the farmers in tropical Asia. It seems probable that a yield level up to four to five tons of paddy per hectare would not pose any particular requirements for physical properties on paddy soils, and yield difference would be caused mostly by differences in chemical soil fertility. In addition, the investment required to improve physical properties to meet the Japanese standard is so forbiddingly high that it is impractical in most areas of tropical Asia.

Water Quality and Soil Fertility

The quality of irrigation water exerts an influence on the fertility of paddy soils. If we assume the water requirement for a single crop of rice to be 1000 mm, only 1 ppm of an element in the irrigation water would supply about 10 kg of that element per hectare per crop. If the water contains 2 ppm of K^+, the total supply of K_2O from the water would be about 25 kg per hectare, which is a figure large enough to be taken into account when one is to apply fertilizers. In a study conducted in Japan by Yoshida (1961), it was shown that the amount of calcium and magnesium supplied by irrigation water exceeds the requirement for rice by two to six times, whereas the supply of silica, potash, phosphorus, and nitrogen satisfies the requirement of the respective elements by 30, 17, 1, and 8 percent. These figures vary widely from one place to another, but they certainly show the importance of irrigation water as a source of SiO_2, Ca^{2+}, Mg^{2+}, and K^+.

In Table 3.3 data of the quality of river water in different countries of tropical Asia are given. The quality of river water is naturally influenced by the geology and degree of weathering in the catchment area. Rivers draining areas of recent volcanic activity are typically rich in silica and moderate to rich in bases, depending on the nature of volcanic ejecta. This is seen in the case of volcanic regions in Java. A relatively high silica content for Japan is also due to a wide occurrence of volcanic ash areas. But sometimes river water from volcanic areas is very acid because of the presence of sulfuric acid. In Japan there are many such instances.

In areas which have undergone intensive weathering and leaching, river water is dilute, as in the cases of West Malaysia and Sri Lanka. As such, the quality of river water is a good indicator of the mineral reserves of the soils in the catchment areas. But even in Malaysia and Sri Lanka, the silica content of river water is moderate to high when compared with the world average, which may be a reflection of tropical weathering, that is, intensive ferrallitization and concurrent desilication.

Floating matter and turbidity indicate the intensity of erosion in the catchment areas and the amount of silt load carried down by the rivers.

The amount of nitrogen (both ammonia and nitrate) and phosphorus supplied by river water is usually negligibly small. In some instances nitrogen supply reaches a toxic level, especially where a river is polluted by urban sewage or by industrial wastes. Rivers that drain swamp areas may have a relatively high nitrogen content, as observed in Malaysia. But usually one cannot expect a supply of nitrogen and phosphorus from river water or silt, and this should be borne in mind in the evaluation of the short-term effect of silting (see chapter 9).

Nitrogen Status

As stated in chapter 2, the ease of maintenance of soil fertility is a remarkable advantage of paddy fields. Low-lying terrain and the associated restricted internal drainage are the two factors that bring about this advantage. But these two factors do not positively contribute to the status of the two very important major elements, nitrogen and phosphorus. Waterlogging and reduction are the processes that are not only specific to the paddy field condition but also play a positive role in the acquisition and release of these two major nutrient elements.

In the waterlogged paddy soils, part of the soil organic matter is decomposed, first aerobically and later anaerobically. Ammoniacal nitrogen thus released from decomposing organic matter during the cropping season is an important source of nitrogen taken up by rice plants. In many studies conducted in Japan, it has been shown that even in a field which has had sufficient fertilizer applied to it, nitrogen supply through mineralization of organic matter is greater than that from fertilizers, by a ratio of about 2:1.

This is true even in the tropics. In a recent study conducted by Koyama *et al.* (1973) in Thailand, it was shown that more than 60 percent of nitrogen taken up by rice plants by the time of harvest comes from mineralization of soil organic matter. By the use of ^{15}N they estimated the amount of mineralized soil nitrogen during the cropping season to be 131 kg/ha in a brackish water alluvial soil near Bangkok (1.4 percent organic carbon and 0.13 percent total nitrogen). The amount estimated was a little lower than that in a Japanese paddy soil with low fertility. They further indicated that the percentage distribution of the mineralized soil nitrogen absorbed by plants was 60 percent during the period up to the end of tillering, 36 percent during the panicle formation stage, and 4 percent during the ripening stage. Comparing these figures with 45:27:28 percent for the respective periods observed for a Japanese soil, they concluded that the high temperature in the tropics is relevant to this pattern of mineralization of soil nitrogen and, accordingly, a topdressing of nitrogen fertilizer at a later growth stage would probably be effective.

The initial release of nitrogen from soil organic matter is known to be enhanced by drying the soil prior to waterlogging. In Japan this is called the "soil-drying effect" and finds application in practice; farmers try to dry soil as much as possible before preparing it for rice

TABLE 3.3   WATER QUALITY OF RIVERS IN TROPICAL ASIA

RIVER	SAMPLING SITES	Ca	Mg	Na	K	HCO$_3$	SO$_4$	Cl	SiO$_2$	RESIDUE ON EVAP.	FLOATING MATTER	pH	HARD-NESS
						mg/l							
	THAILAND[1]												
Mekong	Nong Khai	31.1	5.7	7.7	1.6	115.6	14.7	6.2	15.0	139.2	174.1	6.9	101.0
	Mukdahan	26.8	4.9	7.5	1.4	100.3	12.2	6.6	13.8	124.3	99.9	6.9	87.3
Chi	Khon Kaen	21.1	4.0	57.1	3.9	69.7	5.2	94.5	10.1	244.7	60.5	6.7	69.0
Mun	Ubon Ratchathani	10.9	2.3	40.0	2.8	42.4	2.0	61.6	10.8	165.2	46.6	6.4	36.8
Ping	Chiang Mai	23.1	3.8	3.9	2.7	102.1	0.7	0.5	23.6	108.6	103.1	6.7	73.5
Wang	Lampang	28.3	4.0	5.2	3.6	119.0	1.7	0.8	22.0	125.8	228.9	6.9	87.5
Yom	Sukhothai	31.6	6.0	7.9	2.8	140.8	3.2	0.6	21.1	143.4	296.8	6.8	103.7
Nan	Nan	25.3	3.5	5.8	1.1	105.5	0.3	1.5	20.3	110.8	120.2	6.9	77.7
Chao Phraya	Uthai Thani	13.3	2.7	8.8	4.4	71.6	0.1	6.1	18.2	94.7	50.0	6.6	44.4
	Ayutthaya	22.5	3.9	8.4	3.2	99.7	1.2	7.5	18.3	119.8	192.1	6.7	72.2
Pasak	Saraburi	46.8	4.9	6.7	2.9	164.6	0.3	3.9	19.0	182.3	65.9	6.9	137.0
Mae Klong	Ban Pong	37.7	8.6	3.0	1.9	162.1	0.2	2.1	14.1	150.1	56.9	7.0	129.5
	WEST MALAYSIA[2]												
Kelantan	Pasir Mas, Kelantan	4.7	0.8	2.8	1.3	26.4	1.1	0.1	16.2	42.7	26.3	6.7	
Trengganu	Kuala Trengganu, Tregganu	2.6	0.4	3.4	1.2	15.6	1.3	2.0	13.3	33.3	6.1	6.4	
Pahang Tua	Pekan, Pahang	4.1	0.7	2.9	1.6	22.8	1.6	0	14.8	46.2	56.7	6.3	
Kesang	Muar, Johore	9.9	7.1	11.4	4.5	6.1	102.8	16.5	14.8	221.8	46.9	4.4	
Tengi	Kuala Selangor, Selangor	1.2	0.2	2.3	1.1	0.7	2.7	2.3	5.3	71.4	40.7	4.3	
Muda	Pinang Tunggal, Penang	3.6	0.6	1.9	2.1	19.6	1.8	0.1	13.7	44.2	38.4	6.0	
Padang Terap	Padang Terap, Kedah	4.3	0.9	2.2	1.9	21.2	1.7	1.7	13.0	42.3	51.3	6.1	

		SRI LANKA[3]										
Mahaweli	Peradeniya	4.1	1.3	2.1	1.0	20.8	0.6	2.1	8.1	37.6	101.7	6.2
Ganga	Manampitiya	10.0	3.1	3.7	1.4	49.1	1.2	3.1	15.1	66.2	150.6	6.8
Gal Oya	Inginiyagala	7.3	2.0	3.9	2.2	38.5	0.3	0.9	16.1	58.0	34.0	6.8
Walawe Ganga	Embilipitiya	12.0	2.8	3.9	1.7	55.7	1.1	4.0	16.1	72.0	189.6	6.8
Gin Ganga	Baddegama	2.0	0.6	2.3	0.9	10.6	0.7	2.7	7.9	25.4	30.6	6.4
Kelani Ganga	Hanwella	2.2	0.8	1.8	0.8	13.8	0.6	2.0	7.3	27.3	69.5	6.3
Deduru Oya	Nikaweratiya	13.0	4.3	7.9	1.9	71.4	1.0	5.6	20.9	95.8	101.5	7.0
		CAMBODIA[4]										
Mekong	Phonom Penh	19.4	4.3	6.2	2.1	77.5	8.3	12.7	—	—	—	7.9
		INDONESIA[5]										
Geology of river catchment												
Volcanic		12.2	4.0	10.5	4.1	—	11.4	4.0	42.3			
Marl		101.3	16.3	57.6	5.7	—	131.5	26.8	30.2			
Lime		73.7	8.3	6.6	1.7	—	10.1	3.8	18.2			
Acid tuff loam		6.7	1.9	6.8	3.6	—	6.6	1.4	28.6			
		JAPAN[6]										
Japan average of 225 rivers		8.8	1.9	6.7	1.2	31.0	10.6	5.8	19.0	74.8	29.2	
		WORLD[7]										
World average		15.0	4.1	6.3	2.3	58.4	11.2	7.9	13.1	—	—	

1 Annual average data; Kobayashi, 1958.
2 Annual average data; cited from Kobayashi, J., unpublished.
3 Annual average data; cited from Kobayashi, J., unpublished.
4 Simple averages of the data obtained by the Institute de Pasteur for January–May, 1961–1963, and June–December, 1961–1963; cited from Netherlands Delta Development Team, 1974b.
5 Simple average of unsystematic measurements; Kojima *et al.*, 1962; (originally from van der Giessen, 1943)
6 Kobayashi, 1961
7 Kitano, 1969 (originally from Livingstone, 1963)

cultivation by making ridges or by cutting small ditches. The same effect would occur in paddy soils in tropical Asia, especially in the tropical savanna climate zone, where an extreme drought precedes the rainy season. But the utilization of nitrogen thus released is difficult under the uncontrolled water conditions which prevail in the greater part of tropical Asia.

In reduced conditions in paddy soils during the cropping season, most of the ammoniacal nitrogen released from organic matter stays as it is, and is absorbed on soil exchange complexes, without undergoing nitrification. Thus the amount of nitrogen lost through leaching is lower than that from upland soils under the same terrain conditions.

Although some of the nitrogen taken up by rice plants is returned to the soil, directly as stubble, roots, and straw, and indirectly as cow dung and other animal excreta, the nitrogen status of paddy soils, especially of those receiving no nitrogen fertilizers, as is common in tropical Asia, is inevitably lowered if there is no mechanism to compensate for the loss by cropping. Taking, for example, a Thai paddy soil studied by Koyama *et al.* (1973), we can make the simple calculation that its total nitrogen, 0.13 percent, which is equivalent to 1960 kg/ha of nitrogen in 10 cm of surface soil with a bulk density of 1.3 g/cc, will be used up in about 13 years at a rate of mineralization of 131 kg N per crop per year, if there is no return of utilized nitrogen.

Since other natural sources of nitrogen, such as rainfall and irrigation water, are hardly likely to account for the amount of nitrogen used by the rice crop, it is obvious that microbial nitrogen fixation has made it possible to produce more than one ton of paddy per hectare continuously over hundreds of years under conditions where no methods of replenishing fertility have been adopted.

There are many kinds of nitrogen fixers thriving in waterlogged paddy soils. Matsuguchi and Tangcham (1974) recently made a study on the free-living (nonsymbiotic) nitrogen fixing microorganisms in Thailand and gave the following results, in number of cells per gram of soil; Azotobacter. $0–10^4$, Beijerinckia $0–10$, Clostridia $10^4–10^6$, nonsulfur purple bacteria $10^2–10^5$, and blue green algae $10^2–10^3$. The population density of these nitrogen fixers is dependent on such soil properties as pH, organic matter content, available phosphorus content, and so forth.

The nonsymbiotic, heterotrophic nitrogen fixers, Azotobacters and Clostridia, are not very effective in paddy soils. The former is an aerobe and cannot be very active in the generally reduced paddy soil condition. The latter is a heterotrophic anaerobe, and its population density is usually high, but according to a recent study by Yoshida *et al.* (1973), its role in nitrogen fixation is not significant even in the rhyzosphere where a plentiful supply of energy sources is available.

Photoautotrophs such as blue green algae and photosynthetic bacteria are known to be important in paddy soils. Blue green algae flourish in the surface water and the uppermost part of the plow layer soil. They can assimilate both carbon and nitrogen and thus require no extraneous energy sources. The amount of nitrogen fixation in the surface water of a paddy field, which is probably mostly carried out by blue green algae, is estimated as ranging from 3.2 kg N/ha in the presence of rice plants to 10.9 kg N/ha in their absence (Yoshida *et al.*, 1973).

Photosynthetic bacteria (Res. Group of Soil Microorg., 1966) are able to assimilate atmospheric carbon dioxide, but one of the three species of the group, nonsulfur purple bacteria (Athiorhodaceae), can also utilize organic acids as an energy source in the absence of light. All the photosynthetic bacteria are strict anaerobes and can flourish in the strongly reduced condition of paddy soils during the cropping season. Their nitrogen fixing ability is strongest in anaerobic light conditions, but nonsulfur purple bacteria can fix a fair amount of nitrogen even in the dark, if pyruvic acid is present. As the latter acid is excreted by many bacteria, nonsulfur purple bacteria would contribute to nitrogen fixation through symbiosis with other bacteria. No data are yet available concerning the amount of nitrogen fixed by photosynthetic bacteria in the field condition.

A study by Yoshida (1972) shows that some facultative anaerobes in the rhyzosphere are capable of assimilating nitrogen. Rice can transmit atmospheric nitrogen and oxygen to the rhyzosphere through its gas-conductive organ. The bacteria can assimilate this nitrogen obtaining energy from the decomposition of root exudates and root debris.

Estimates of total nitrogen fixation by Matsuguchi and Tancham (1974) range from 0.5 to 54 kg N/ha for different soils in Thailand. But in more than half, it was less than 5 kg N/ha. However, Yoshida's (1971) result, 79.8 kg N/ha in planted submerged soil and 42.5 kg N/ha in unplanted submerged soil in comparison with 5.4 kg N/ha in planted upland soil and 2.7 kg N/ha in unplanted upland soil, clearly shows the much higher efficiency of nitrogen fixation in paddy soils.

The effects of liming and phosphorus application in stimulating microbial nitrogen fixation have been ascertained. Liming in paddy fields has been practiced by Japanese farmers from early times, not to correct soil acidity but to stimulate microbial decomposition of soil organic matter, and also, although it was not realized, enhancing microbial nitrogen fixation. One long-term experiment conducted at the Fukui Agricultural Experimental Station (Ueda *et al.*, 1975) clearly shows the positive effect of liming in increasing the rice yield.

We have so far discussed the positive aspect of nitrogen economy in paddy soils. The negative aspect is the denitrification process. For this process differentiation of a thin ($<$ 1 cm) surface oxidative layer due to oxygen diffusion through shallow surface water is assumed. As stated earlier, nitrification in paddy soils during the cropping season is generally restricted. But once such layer differentiation occurs, nitrifiers are sufficiently active in the thin oxidative layer for nitrification of applied or evolved ammoniacal nitrogen to take place. Because it is an anion, the nitrate thus formed is easily diffused into the reductive layer where it is reduced to gaseous nitrogen by denitrifiers. Denitrifiers are highly active only at a relatively shallow depth ($<$ 5 cm) (Yoshida and Padre, 1974).

The amount of nitrogen lost from paddy soils by denitrification is estimated as considerably large. The applied ammoniacal nitrogen recovery rate does not usually exceed 50 percent and is often around 30 percent (Sanchez, 1972). Part of the remainder may have been temporarily immobilized by microbes, but the greater part is lost by denitrification. In a laboratory experiment with tagged nitrogen, Yoshida and Padre (1974) recovered only 46 percent of the applied nitrogen after a five-week incubation period. In their study the use of nitrification inhibitors contributed to reduction of denitrification loss during the same period. It is also an established fact that deep placement of N-fertilizers reduces denitrification (Broeshart, 1971).

As denitrification is dependent on the redox regime of the soil, the actual extent of the denitrification loss of nitrogen is determined by terrain and soil conditions specific to a location; a deeply and continuously flooded paddy soil with strongly reductive condition may not show as much denitrification loss as in a shallowly flooded and better-drained soil. Thus, further field studies, such as the one conducted by Araragi and Tangcham (1974), must be carried out in the tropical Asian condition.

Phosphorus Status

Sufficient phosphorus is not usually supplied by irrigation water, nor by microbial fixation. Therefore, the phosphorus status is primarily determined by the total phosphorus content of the parent material. But even when the total phosphorus level is the same, the availability of phosphorus can vary greatly from one soil to another. The waterlogging of paddy fields exerts a great effect on this phosphorus availability.

First, we will show the relative advantage of paddy soils over upland soils with respect to phosphorus. Kawasaki (1953) summarized the results of fertilizer trials, both in pots and in fields, conducted at many different places throughout Japan, as shown in Table 3.4. The table shows index numbers for the yield of rice, upland rice and upland cereals, including wheat, barley, naked barley, and two-row barley, taking the yield of the complete plot as 100. No-P plots of rice show indices of 89 for pot, and 95 for field trials, while the yield of No-P plots of upland rice falls to 79 and 84, respectively, for pot and field conditions. The upland cereals, wheat, and barleys, are even more susceptible to the lack of phosphorus, with yield indices of only 62 and 69. The number of experiments suggests that this difference between types of crop is significant. Lowland rice is in an advantageous position relative to upland rice and wheat and barleys, and this is certainly brought about by the practice of waterlogging paddy fields.

The enhanced availability of phosphorus under submerged conditions is usually attributed to (Patrick and Mahapatra, 1968):

(1) reduction of insoluble ferric phosphate to more soluble ferrous phosphate,
(2) release of occluded phosphate by reduction of hydrated ferric oxide coating,
(3) displacement of phosphate from ferric and aluminum phosphates by organic anions,
(4) hydrolysis of ferric and aluminum phosphates, and
(5) anion (phosphate) exchange between clay and organic anions.

The first two mechanisms appear to be most significant. Patrick (1964) showed a marked increase in extractable phosphorus with the lowering of redox potential below 200 mV, at which Eh level ferric ion also began to be reduced to the ferrous form. High content of Fe-bound phosphorus

TABLE 3.4   RESULTS OF FERTILIZER TRIALS IN JAPAN

Crop	Trial	Complete	No-N	No-P_2O_5	No-K_2O	None	No. of Trials
Rice	Pot	100	55	89	88	53	2850–2898
	Field	100	83	95	96	78	1161–1187
Upland rice	Pot	100	49	79	92	41	176
	Field	100	51	84	75	38	117–126
Upland cereals*	Pot	100	34	62	81	26	1937–1991
	Field	100	50	69	78	39	822–841

*Such crops as barley, naked barley, and wheat are included.
Figures are index numbers.
From Kawasaki (1953).

and occluded phosphorus in paddy soils has been shown to exist by Cholitkul and Tyner (1971), using the Chang-and-Jackson fractionation method. Our previous results (Kawaguchi and Kyuma, 1969a) showed that many Thai soils contain a high proportion (66–80 percent) of total fractionated phosphorus in these two forms, irrespective of the parent material and age of the soils.

The third mechanism is often mentioned in relation to the chelating ability of organic anions, but it is not particularly specific to waterlogged conditions, except that in reduced conditions many kinds of partially decomposed organic matter that have chelating ability are produced. The fourth mechanism is related to the pH increase that accompanies the reduction process. The fifth is somewhat ambiguous, but if it means the release of phosphates by anion exchange on clay, again pH increase should be relevant.

One additional mechanism operative in a strongly reductive condition is expressed by the following equation (Ojima and Kawaguchi, 1959):

$$Fe_3(PO_4)_2 + 3H_2S \rightarrow 3FeS + 2H_3PO_4$$

Organic phosphates, which constitute a high percentage of total phosphorus in some paddy soils, are said to contribute significantly to the increase in solubility of phosphorus. Furukawa and Kawaguchi (1969) reported that iron phytates are particularly important in this mechanism.

These various findings illustrate only the relative advantage for paddy soils in phosphorus economy in comparison with upland soils. This does not exclude the possibility of phosphorus deficiencies in the paddy soils of tropical Asia (see chapter 9). An experiment carried out over a 38-year period at Hiroshima Agricultural Experimental Station in Japan showed that the phosphorus content of the No-P plot fell to as little as 38 mg P_2O_5/100 g, but the rice yield stayed at a level 90–95 percent that of the continuous-P plot (Komoto *et al.*, 1968). Shiga Agricultural Experimental Station (1963) carried out a similar experiment. In this case the No-P plot showed about a 10 percent decrease in rice yield even in the earliest years of the experiment, when the total phosphorus level was around 70 mg/100g as P_2O_5. A marked decrease in rice yield set in only after the total phosphorus level of the surface soil fell below 31–32 mg/100 g. The soils of both stations are sandy loams to loams of granite origin, and the relative proportion of available phosphorus in the total is considered high.

There is another example from tropical Asia. One of our samples, Ca-3 from Battambang, Cambodia, contained 44 mg P_2O_5/100g, but in field trials it was shown to be severely deficient in phosphorus, and rice responded noticeably to phosphate application. (S. Hirano, personal communication.) The soil in this case is a kaolinitic heavy clay.

As illustrated by these examples, it is difficult to set a generally applicable threshold value of total phosphorus content, but it may be possible to say that soils containing less than 40 mg total P_2O_5/100g of soil are liable to be deficient in phosphorus even at the present low, general yield level in tropical Asia.

4

Samples and Methods

Considering the great importance of paddy soils in tropical Asia as a medium of food production for so many people, there have been relatively few publications describing the nature and properties of these soils. There is more or less comprehensive treatment of paddy soils in Thailand (Sarot Montrakun, 1964; Kawaguchi and Kyuma, 1969a, Motomura, 1973; Prachak Charoen, 1974), but there are very few publications for other countries; we only know of Kawaguchi and Kyuma (1969b) for West Malaysia and Panabokke and Nagarajah (1964) for Sri Lanka. As yet there is no treatise that covers a wider region on a comparative basis.

Therefore, since 1963, we have conducted studies to fill this gap. The methods adopted and the samples used are described below.

4.1 SAMPLING PROCEDURES

As stated in chapter 1, a total of some 83 million hectares of paddy land is cultivated annually for rice in tropical Asia. Distribution of this land is shown in Table 1.6. Because the area for double cropping is very small, the figures quoted in the table may be regarded as the total extent of paddy soils in tropical Asia.

Ideally, the samples for this study should have represented the different kinds of paddy soil in proportion to their distribution in the region. Obviously, this was not possible in our study because of the lack of information. An alternative procedure is random sampling, taking a certain number of samples from a certain area of paddy land. This may be possible for a government agency but not for individual researchers like ourselves.

The samples we took are not representative of tropical Asia as a whole. First of all, due to limitations in budget, manpower, and so forth, not all the countries in tropical Asia could be covered. Paddy soils in Pakistan, Nepal, Vietnam, Laos, East Malaysia, and Brunei were not included in this study. (Note: The Mekong delta of Vietnam was surveyed after the preparation of the first draft of this book. The results of the study are included in chapter 8.) In the actual sampling we paid more attention to taking samples that could represent each of the countries covered. But even this presented difficulties. The sampling procedures were subject to many limitations, the more important ones being:

(1) Availability of previous data on the distribution of different kinds of paddy soil. Number and kind of samples to be taken in each area and approximate sampling sites were determined after consultation with local soil specialists. Quantity and quality of available data, however, varied greatly from one country to another, and even from one region to another within a country. For instance, in Thailand and Bangladesh, FAO soil survey projects were in progress at the time of our sampling, and we received much assistance from the FAO and local soil survey specialists during our surveys and sampling. But even within Thailand, not much data were available for the Upper Central Plain, and we had to collect samples based on the extent of rice land as could be read from topographical maps and the physiography as observed in the field. Naturally the samples thus collected are less satisfactory than those sampled on the basis of previous soil surveys.
(2) Accessibility to the site. To save time, and sometimes for reasons of personal security, we had to confine our sampling operation to areas with good accessibility. This usually meant the areas along the better maintained roads.
(3) Areas surveyed. In some countries, particularly India, Indonesia, the Philippines, and Burma, only part of the country could be sampled. The Burmese samples were collected only for a preliminary idea of their nature, on a short trip designed for other purposes, and were thus most unsatisfactory for the purposes of this study.

The number of samples collected and the areas covered in each country are as follows:

Bangladesh	53	throughout the country
Burma	16	part of the Irrawaddi Delta and Upper Burma
Cambodia	16	mainly areas around Grand Lac
Indonesia	44	Java only
Malaysia	41	throughout West Malaysia
Philippines	54	Luzon, Panay, Leyte, and Mindanao
Sri Lanka	33	throughout the country
Thailand	80	throughout the country

It is obvious from this that the number of samples taken is not in proportion to the extent of rice land in each country. This disparity in the number of samples from different countries is the greatest drawback of the study.

In spite of this, the samples from all the countries, except Burma, may be regarded as reasonably representative of the respective countries, or at least the major rice-growing areas thereof. The total number of samples, 410, is substantial enough to enable us to make some generalizations about paddy soils in tropical Asia, especially with respect to their material features and fertility status.

The location, parent material, and local soil classification of the samples, by country, are given in Appendix 1 and maps of the locations are given in chapter 8.

4.2 METHOD OF FIELD INVESTIGATION

Landform and parent material were taken into consideration in deciding the sampling site. At most of the sampling sites, except in the case of the Burmese samples, a pit, one meter deep, was dug to show soil profiles. Auger boring replaced pit digging at some sites. In the case of the Burmese samples only the surface soil was scooped. Most of the description is according to the soil survey method used in Japan. Though the Japanese method is a modification of the one described in the Soil Survey Manual of USDA (1951), its major deviation from the latter is in field texture grading. We use international size limits for the different textural separates, and the classification system proposed by Tommerup (1934) for international use.

In identifying and naming soil horizons special attention was paid to illuvial iron and manganese accumulation below the surface plow layer in a search for the superimposed morphology under paddy rice cultivation. As stated in chapter 3, such profiles were rare and most of the paddy soils investigated had an Apg-Cg horizon sequence.

Information obtained from the profile investigation, however, was not included in this book, except for the general characteristics already described in chapter 3. This is because (a) paddy soils have relatively little profile development, and (b) subsoil characters have little effect on soil capability in terms of crop performance (see the discussion on physical properties of paddy soils, chapter 3).

4.3 METHODS OF LABORATORY STUDIES

All the soil samples were taken back to the Laboratory of Soils in Kyoto University without undergoing quarantine, by special permission of the Plant Protection Office of the Ministry of Agriculture and Forestry. They were air-dried, and prepared as air-dry fine soil by mechanical or hand crushing and sieving through a 2 mm round-hole sieve. No special consideration was paid to acid sulfate soils, because samples inevitably underwent some alterations during the period of transportation (about two to three months).

Items for analysis are listed below:

  (1) pH
  (2) Total carbon
  (3) Total nitrogen
  (4) Ammoniacal nitrogen
      (by anaerobic incubation)
  (5) Total phosphorus
  (6) Available phosphorus
      (Bray-Kurtz No. 2 and 0.2N HCl soluble)
  (7) Cation exchange capacity
  (8) Exchangeable
      cations (Ca^{2+}, Mg^{2+}, Na^+, and K^+)
  (9) Available silica
 (10) Mechanical analysis

Details of the methods adopted for these items are given in Appendix 2. When necessary, such items as electric conductivity, free iron and manganese oxides, and phosphorus absorption capacity, were also analyzed.

In addition to these routine analyses, clay mineralogical analysis and total chemical analysis were carried out as described in chapter 5, section 2.

In order to secure a specified level of precision and accuracy, the following measures were taken: Prior to routine analysis, the method to be used was examined in experiments designed for multifactorial statistical analysis, and then standardized in order to minimize errors, costs and labor (Matsuo, 1968). In routine analysis, precision and accuracy were checked by means of $\bar{x}$-R control charts with respect to standard solutions and standard samples, which were normally one per twenty samples.

4.4 METHODS OF DATA PROCESSING

A soil may be characterized in many ways. Terrain and profile descriptions are ways of characterizing the soil in the field, from which qualitative and a few quantitative data are obtained. Laboratory analysis is another way of characterizing the soil from which mostly quantitative data are acquired. Thus there are usually many data by which one soil may be described and characterized.

Furthermore in a study such as this one, the number of soils studied is very large, and for each of these soils many data are produced. The total number of data can easily amount to tens of thousands, and such data are of wide variety, measured in a nominal, ordinal, or interval scale.

Some sort of processing is necessary before one can adequately summarize the many data and draw useful conclusions from them. Multivariate statistical methods, making use of the electronic computer, have been developed in recent years to cope with this situation. A variety of methods is available today to handle many kinds of data and problems. They are classified as in the following table (after Akuto, 1971):

External Criterion Variable and Its Nature		Explanatory Variable	Method
Present	Quantitative (estimation of a quantity)	Quantitative	Multiple regression analysis
		Quantitative and/or Qualitative	Hayashi's theory of quantification No. 1
	Qualitative (estimation of a quality)	Quantitative	Discriminant function
		Quantitative and/or Qualitative	Hayashi's theory of quantification No. 2
Absent (classification)		Quantitative	Principal component analysis
			Factor analysis
		Quantitative and/or	Cluster analysis
		Qualitative	Hayashi's theory of quantification No. 3

In this study we adopted many such multivariate methods. The following is a list of the methods used:

Multiple regression analysis	in chapter 5
Hayashi's theory of quantification No. 1	in chapter 5
Cluster analysis (numerical taxonomy)	in chapter 6
Discriminant function	in chapter 6
Principal component analysis	in chapter 6 and 7
Factor analysis	in chapter 7

Each of the methods used is briefly explained in the relevant chapters. For details of the methods reference should be made to the appropriate books, for example, Okuno *et al.* (1971) and Hayashi *et al.* (1970).

The computer used throughout this study is a FACOM-230 series (Fuji Communication and Electronics Co.) at the Data Processing Center of Kyoto University. The computer programs used are mainly those contained in the SPSS (Statistical Package for Social Sciences), Kyoto version (Miyake, 1973), at the Data Processing Center.

5

Description and Correlation Analysis of the Soil Data Obtained for Paddy Soils in Tropical Asia

In this and the following chapters the results of our own studies on paddy-soils in tropical Asia are presented. First, the characters of the soils are described in terms of their mean, mode, and range of variation. Important correlations between these characters are also discussed in this chapter to form the basis for the processing of soil data to be dealt with in the following chapters.

5.1 DESCRIPTION OF FERTILITY CHARACTERS

Analytical data for the 410 surface soil samples are presented and briefly discussed. Histograms are prepared by establishing 10 classes; the class limits are set up (1) to give the same interval to each class, except the two outermost classes, and (2) to put a small percentage (not more than 10 percent) of the entire samples into the uppermost class. For each item, a histogram and/or a table showing country means and other statistics is given. For the purpose of comparison, data of Japanese and Mediterranean (Italy, Spain, and Portugal) paddy soils are also cited in the table (Kyuma *et al.*, 1974). These samples were collected and analyzed by us and can be regarded as representative of the respective regions.

pH (Fig. 5.1. and Table 5.1.)

The single analytical item that is most frequently quoted is pH. As will be seen later, it has many covarying characters, such as base status and clay mineralogy.

The overall mean for the 410 samples is 6.0, but the mode is class 4, which corresponds to pH 5.0–5.5; the distribution is slightly positively skewed. By country, West Malaysian paddy soils have the lowest mean

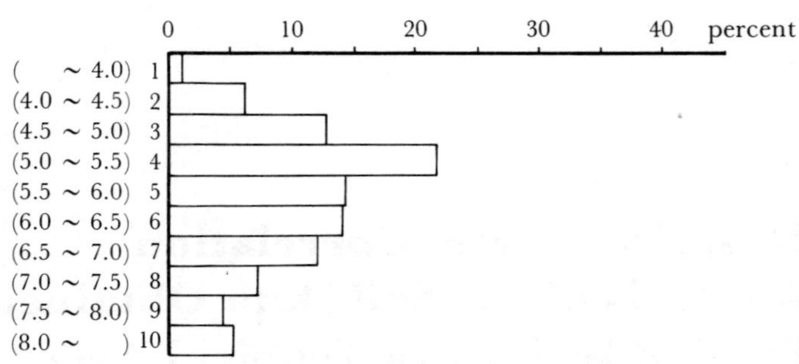

FIG. 5.1   HISTOGRAM FOR pH OF PADDY SOILS IN TROPICAL ASIA

TABLE 5.1   MEAN, STANDARD DEVIATION, MINIMUM AND MAXIMUM VALUES OF pH

COUNTRY	NO. OF SAMPLES	MEAN	STANDARD DEVIATION	MINIMUM	MAXIMUM
Tropical Asia	410	6.0	1.1	3.4	9.3
Bangladesh	53	6.1	1.0	4.3	8.3
Burma	16	6.4	1.3	4.8	8.5
Cambodia	16	5.2	0.8	4.1	7.1
India	73	7.0	1.1	5.0	9.3
Indonesia	44	6.6	0.9	5.1	8.6
W. Malaysia	41	4.7	0.5	3.4	6.1
Philippines	54	6.4	0.6	4.8	8.5
Sri Lanka	33	5.9	0.8	4.9	8.4
Thailand	80	5.2	0.6	3.9	7.5
Mediter. Countries	62	6.8	1.0	5.4	8.6
Japan	84	5.4	0.5	3.8	6.3

value of 4.7, followed by Cambodia and Thailand with 5.2. The acid nature of Malaysian paddy soils has also been observed by other researchers (Arnott, 1964). This is partly explained by intense leaching under the fairly uniformly humid climate and partly by frequent occurrence of paddy soils reclaimed from peaty swamps, which have a high amount of unsaturated organic matter. The low pH of Cambodian soils is mainly due to the abundance of sandy, acidic parent materials that have undergone severe weathering and leaching. The same applies to the soils of Northeast Thailand (Khorat Plateau).

Another cause of the acidity of Thai paddy soils is the wide distribution of acid sulfate soils and the related acid soils in the Bangkok Plain.

The rest of the countries have high mean pH values nearly equal to or higher than 6.0, which is much higher than the mean pH value of 5.4 for Japanese paddy soils. Concentration of rainfall in half a year, resulting in a large runoff relative to percolation, and alternation of rainy and dry seasons are the factors that explain the difference in the mean pH values between Japan and, for example, Sri Lanka which has even greater annual precipitation than Japan.

The number of soils with pH values higher than 7 in each country is as follows:

Bangladesh	10	Malaysia	0
Burma	7	Philippines	4
Cambodia	1	Sri Lanka	4
India	30	Thailand	2
Indonesia	12		

The occurrence of the high pH soils is associated with the climate and/or parent material. Climatic influence is obvious from the fact that all four samples from Sri Lanka occur in the Dry Zone, all twelve soils in Indonesia come from Central and East Java where the climate tends to be drier than in West Java, and all seven Burmese soils are from Upper Burma where the climate is distinctly drier than that of Lower Burma.

The effect of parent material is best seen in Bangladesh, where climate is almost uniform throughout the country. All the high pH soils occur in the region covered by recent Gangetic alluvia. The calcareous nature of the Gangetic alluvia can be traced back to northern India. Local calcareous materials, such as weathering products of limestones and basic igneous rocks, can produce small patches of high pH soils typically grumusolic in nature (Cambodia and Thailand). Another source of high pH soils is deposits in depressions in the wet and dry climate. Some of the high pH soils from India and the Philippines fall into this category.

Among the high pH soils from India there are two soils having pH values above 9. Both of these alkaline soils have high (more than 15 percent) exchangeable sodium.

Soils with exceptionally low pH, below 4.5, are also present, three from Cambodia, fifteen from Malaysia, and five from Thailand. Many of these are peaty and/or swampy with sediments of brackish environment. Among the samples collected, six are acid sulfate soils, all of which have pH values below 4.5. Of these, two Malaysian soils and one Thai soil have a pH even lower than 4. During the period of waterlogging for rice cultivation, soil pH tends to rise and therefore even these acid sulfate soils are usually tolerated by rice plants.

Total Organic Carbon (TC) (Fig. 5.2 and Table 5.2)

The histogram is strongly positively skewed. More than 55 percent of the total samples fall into classes 2 and 3, corresponding to 0.5–1.5 percent organic carbon. The overall mean is 1.4 percent, which is much lower than in middle latitude countries. The most plausible explanation

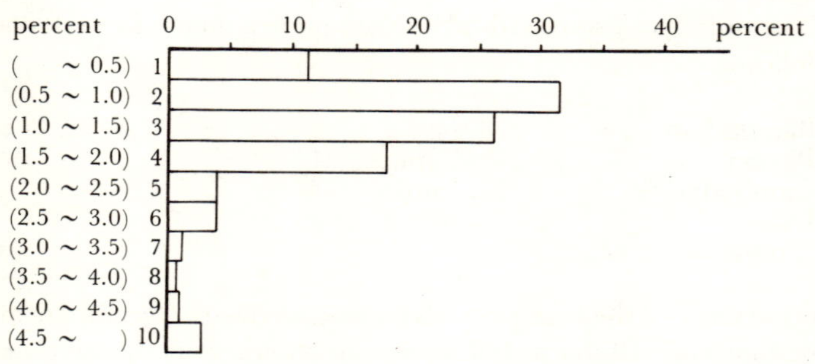

FIG. 5.2   HISTOGRAM FOR TOTAL ORGANIC CARBON CONTENT OF PADDY SOILS IN TROPICAL ASIA

TABLE 5.2   MEAN, STANDARD DEVIATION, MINIMUM AND MAXIMUM VALUES OF TOTAL ORGANIC CARBON
(as percentage of air-dry soil)

COUNTRY	NO. OF SAMPLES	MEAN	STANDARD DEVIATION	MINIMUM	MAXIMUM
Tropical Asia	410	1.41	1.28	0.12	11.40
Bangladesh	53	1.18	0.83	0.47	6.00
Burma	16	1.21	0.50	0.37	2.32
Cambodia	16	1.09	0.77	0.24	2.88
India	73	0.85	0.37	0.28	1.90
Indonesia	44	1.39	0.76	0.50	5.60
W. Malaysia	41	3.36	2.53	0.60	11.40
Philippines	54	1.66	0.64	0.52	3.30
Sri Lanka	33	1.41	1.50	0.18	8.49
Thailand	80	1.05	0.67	0.12	2.95
Mediter. Countries	62	1.82	1.45	0.35	8.60
Japan	84	3.33	2.02	1.00	11.36

for this is high annual temperature and, accordingly, a high turnover rate of organic residues.

West Malaysia has an exceptionally high mean TC value. Of the eleven samples falling into class 10 of the histogram, eight come from Malaysia. These high TC samples from Malaysia are from swampy lowlands with peaty organic matter, which has been mentioned in relation to pH. Even after excluding such swampy soils, however, Malaysian paddy soils have the highest mean TC content of 2.0 percent. Soils from the Philippines, Sri Lanka, and Indonesia form the second group. Internal heterogeneity, however, can be noted for these countries; samples from Mindanao, Leyte, and southern Luzon (Bicol peninsula) in the Philippines, those from Wet and Intermediate Zones in Sri Lanka, and samples from West Java in Indonesia seem to have elevated the country means. The common climatic element for these regions and Malaysia is the absence or shortness of the dry period. Thus, a relatively more humid climate contributes to the accumulation of TC in paddy soils.

The lowest mean TC content is seen for Indian paddy soils for which the coefficient of variation is also the lowest. In accordance with the preceding, a regional climate characterized by very high temperatures at some time of the year and relatively low annual precipitation, appears to be the factor contributing most to the low TC level. Even the maximum figure is less than 2 percent for India. Cambodian and Thai paddy soils have the second lowest mean values.

The maximum TC figures for Sri Lanka and Bangladesh soils are those for swampy soils, whereas that for Indonesian soils is for an ando soil of volcanic ash origin. The high organic carbon content of ando soils is an already established fact in Japan. As the allophanic nature of clay minerals of the ando soil of Indonesia has also been confirmed by us (Kitagawa *et al.*, 1973), we can assume that the mechanism of organic matter accumulation proposed for Japanese ando soils applies to the Indonesian soil.

The minimum figure for Thailand is the overall minimum and is for a sample from the Khorat Plateau.

Total Nitrogen (TN) (Fig. 5.3 and Table 5.3)

The histogram shows a remarkable similarity to that for TC. An even higher concentration of samples (about 65 percent) is seen in classes 2 and 3, which correspond to a TN content of 0.05 to 0.15 percent, and there are fewer samples in the higher TN classes. This fact may be interpreted as indicating that the accumulation of organic

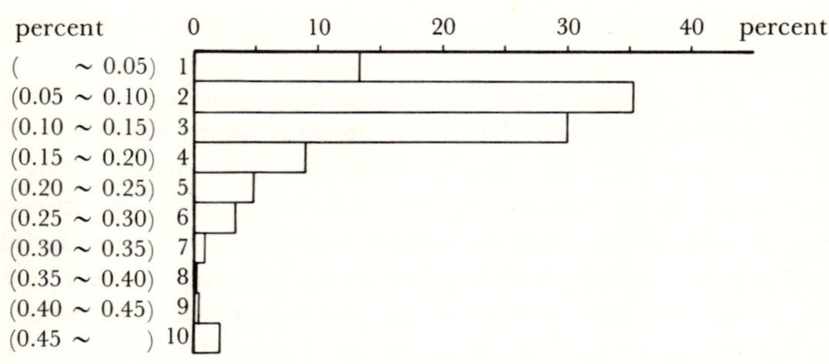

Fig. 5.3 Histogram for Total Nitrogen Content of Paddy Soils in Tropical Asia

Table 5.3   Mean, Standard Deviation, Minimum and Maximum Values of Total Nitrogen
(as percentage of air-dry soil)

Country	No. of Samples	Mean	Standard Deviation	Minimum	Maximum
Tropical Asia	410	0.13	0.11	0.02	0.92
Bangladesh	53	0.13	0.07	0.05	0.52
Burma	16	0.10	0.04	0.04	0.18
Cambodia	16	0.10	0.07	0.03	0.29
India	73	0.08	0.03	0.02	0.20
Indonesia	44	0.12	0.08	0.05	0.59
W. Malaysia	41	0.28	0.21	0.05	0.92
Philippines	54	0.15	0.07	0.06	0.35
Sri Lanka	33	0.13	0.11	0.03	0.64
Thailand	80	0.09	0.06	0.02	0.27
Mediter. Countries	62	0.16	0.09	0.05	0.62
Japan	84	0.29	0.15	0.09	0.91

matter often takes place in the form of less decomposed (or less humified) organic debris having a high C/N ratio.

Again Indian soils contain the least amount of nitrogen, about a quarter of the Japanese mean value. Soils from Cambodia and Thailand have means less than 0.1 percent. Burmese soils also have a low mean. Malaysian soil again have an exceptionally high mean value.

Climate and local relief appear to have contributed the most to the accumulation of both TC and TN. Texture may also have some effect. Correlation and regression analyses of these factors and organic matter are given in the section titled "Effect of Climate, Relief, and Soil Texture on Organic Matter Status."

Carbon/Nitrogen Ratio (C/N Ratio) (Fig. 5.4 and Table 5.4)

As seen from the general parallelism between TC and TN, C/N ratio fluctuates in a narrow range. Mean values of 8.7 and 12.6 for Bangladesh and Indonesian soils, respectively, represent the lower and upper extremes, and all others fall between 10 and 12. The overall mean is 11.2, which is comparable to the Japanese and Mediterranean means.

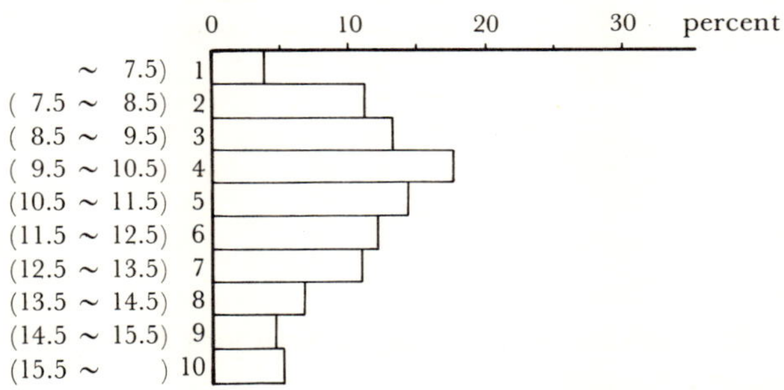

FIG. 5.4   HISTOGRAM FOR CARBON/NITROGEN RATIO OF PADDY SOILS IN TROPICAL ASIA

TABLE 5.4   MEAN, STANDARD DEVIATION, MINIMUM AND MAXIMUM VALUES OF CARBON/NITROGEN RATIO

COUNTRY	NO. OF SAMPLES	MEAN	STANDARD DEVIATION	MINIMUM	MAXIMUM
Tropical Asia	410	11.2	2.7	4.8	25.8
Bangladesh	53	8.7	1.1	7.1	12.4
Burma	16	11.5	1.5	8.6	13.4
Cambodia	16	10.7	1.8	7.7	13.7
India	73	11.7	2.5	4.8	18.4
Indonesia	44	12.6	3.5	8.3	25.8
W. Malaysia	41	11.8	2.3	9.2	19.8
Philippines	54	11.7	2.6	7.7	19.0
Sri Lanka	33	10.5	2.2	6.4	15.6
Thailand	80	11.3	2.9	6.0	22.2
Mediter. Countries	62	10.6	2.5	7.0	25.1
Japan	84	11.6	3.0	3.6	23.2

From this we can conclude that no significant difference is found in the nature of soil organic matter between tropical and temperate paddy soils.

Ammoniacal Nitrogen (NH₃-N) (Fig. 5.5 and Table 5.5 and 5.6)

The overall mean of ammoniacal nitrogen mineralized during a reductive incubation period of two weeks is 8.5 mg/100 g soil, and the

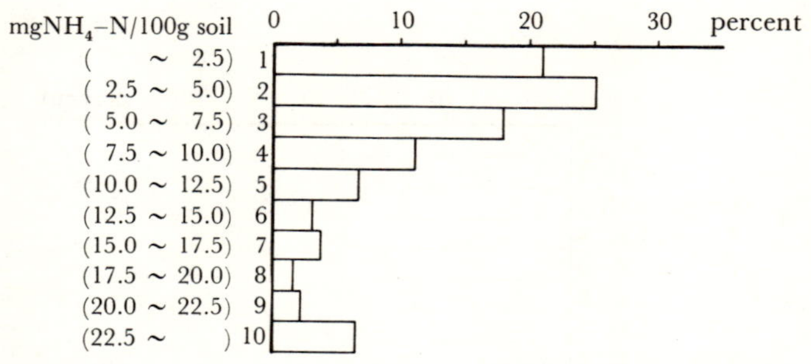

FIG. 5.5   HISTOGRAM FOR AMMONIACAL NITROGEN PRODUCTION OF PADDY SOILS IN TROPICAL ASIA

TABLE 5.5   MEAN, STANDARD DEVIATION, MINIMUM AND MAXIMUM VALUES OF AMMONIACAL NITROGEN PRODUCTION (in mg NH₃-N/100 g air-dry soil)

COUNTRY	NO. OF SAMPLES	MEAN	STANDARD DEVIATION	MINIMUM	MAXIMUM
Tropical Asia	410	8.5	9.3	0.3	63.0·
Bangladesh	53	6.1	4.8	1.6	31.8
Burma	16	2.0	1.6	0.4	6.0
Cambodia	16	4.0	2.2	0.5	7.8
India	73	2.7	1.4	0.5	7.3
Indonesia	44	14.1	8.7	3.6	36.6
W. Malaysia	41	14.9	9.9	3.8	40.3
Philippines	54	17.2	15.7	3.3	63.0
Sri Lanka	33	8.4	6.2	1.1	21.9
Thailand	80	5.2	3.2	0.3	16.4
Mediter. Countries	62	7.6	4.0	1.1	23.6
Japan	84	17.5	7.1	0.6	37.0

TABLE 5.6 MEAN, STANDARD DEVIATION, MINIMUM AND
MAXIMUM VALUES OF PERCENTAGE AMMONIFICATION
(as percentage of total nitrogen)

COUNTRY	NO. OF SAMPLES	MEAN	STANDARD DEVIATION	MINIMUM	MAXIMUM
Tropical Asia	410	6.8	4.7	0.3	26.5
Bangladesh	53	4.5	1.9	1.6	10.8
Burma	16	1.8	1.2	0.3	5.0
Cambodia	16	5.1	2.9	0.7	10.0
India	73	3.9	2.2	1.0	14.8
Indonesia	44	12.4	5.0	4.4	25.4
W. Malaysia	41	6.3	3.6	1.5	14.7
Philippines	54	10.8	6.0	2.8	26.5
Sri Lanka	33	7.0	3.6	1.8	15.7
Thailand	80	6.6	3.6	0.8	18.3
Mediter. Countries	62	4.9	1.9	0.7	9.0
Japan	84	6.5	2.3	0.7	11.8

mean percentage ammonification is 6.8 percent. The variance of the data of NH_3-N production is very large and sample values range from 0.3 mg to 63 mg/100 g soil.

Characteristically low values of ammoniacal nitrogen released and percentage ammonification are seen for Indian and Burmese soils. The sample values of NH_3-N are confined to the first three classes in both Burmese and Indian soils. The low percentage of ammonification may be owing to a high resistance to microbial decomposition of the small reserve of organic matter in these soils.

The high mean for NH_3-N production in Malaysian soils is explained by the high mean TN content. The value of percentage ammonification is just normal for Malaysian soils. But the high mean values of ammoniacal nitrogen for Philippine and Indonesian soils seem mainly because of high percentages of ammonification, averaging 11 and 12 percent, respectively. Reasons for enhanced ammonification cannot be readily given, but biological activity as controlled by base and other nutrient status certainly has some relevance. In Indonesian and Philippine soils these factors may be favorable to ammonification.

Extraordinarily high ammoniacal nitrogen production, that is more than 25 mg/100 g soil, occurs in the samples that were swampy or continuously wet in the field. The drastic change of equilibrium might have caused instability of organic residues that had been accumulated in the swampy condition.

Total Phosphorus (TP) (Fig. 5.6 and Table 5.7)

As phosphorus is readily bound by such soil constituents as iron, aluminum, and calcium to form difficultly soluble phosphate compounds, the level of TP can be modified rather easily by fertilizer applications. This effect can be seen in the case of Japanese paddy soils, whose 220 mg P_2O_5/100 g soil is presumably a result of fertilization. In comparison with this value the mean TP levels of tropical Asian paddy

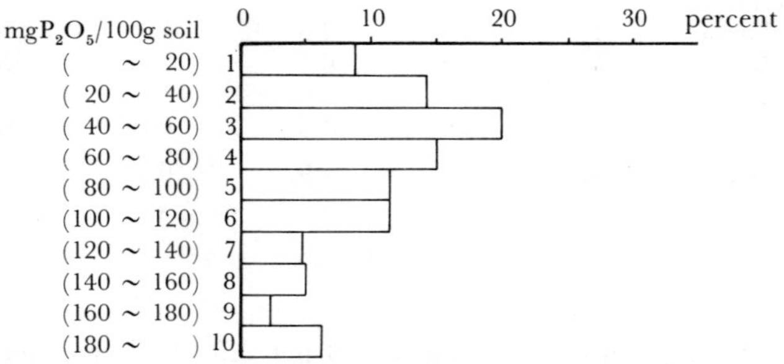

FIG. 5.6   HISTOGRAM FOR TOTAL PHOSPHORUS CONTENT OF PADDY SOILS IN TROPICAL ASIA

TABLE 5.7   MEAN, STANDARD DEVIATION, MINIMUM AND MAXIMUM VALUES OF TOTAL PHOSPHORUS (in mg P_2O_5/100 g air-dry soil)

Country	No. of Samples	Mean	Standard Deviation	Minimum	Maximum
Tropical Asia	410	83.7	66.8	1.0	553.0
Bangladesh	53	84.1	33.4	33.0	160.0
Burma	16	74.4	70.8	9.0	284.0
Cambodia	16	44.3	38.6	5.0	131.0
India	73	92.0	64.5	24.0	475.0
Indonesia	44	133.7	115.6	16.0	553.0
W. Malaysia	41	80.3	42.8	5.0	206.0
Philippines	54	113.6	60.2	23.0	249.0
Sri Lanka	33	72.5	47.1	13.0	174.0
Thailand	80	44.2	42.3	1.0	237.0
Mediter. Countries	62	103.8	46.3	14.0	253.0
Japan	84	220.4	126.9	86.0	861.3

soils are low, indicating little human interference. The overall mean is 84 mg with considerable variance. A high mean TP level of over 100 mg is seen for Indonesian and Philippine soils, whereas a very low level prevails in Cambodian and Thai soils, so that the means are only slightly more than 40 mg.

The number of samples, from each country, falling into class 10 of the histogram that corresponds to more than 180 mg TP/100 g soil, is listed below:

Bangladesh	0	Malaysia	1
Burma	1	Philippines	8
Cambodia	0	Sri Lanka	0
India	4	Thailand	1
Indonesia	10		

The overall maximum TP of 553 mg is contained in an ando soil from Indonesia. From our experience with Japanese ando soils, it may be inferred that the greater part of the phosphorus would exist in organic forms. Seven of ten high TP soils from Indonesia are latosols by Indonesian classification, which would correspond to Oxic Dystropept or Oxic Eutropept of the U.S. soil classification system. These latosols are characterized by relatively low available phosphorus, in spite of their high TP level. This is a reason for the low mean available phosphorus content of Indonesian paddy soils. The high phosphorus content of the Philippine soils seems to stem from parent materials of volcanic origin. Some of the swamp soils are rich in TP, as seen in the examples from Malaysia and Thailand.

The number of soils containing less than 20 mg TP in each country is as follows:

Bangladesh	0	Malaysia	2
Burma	2	Philippines	0
Cambodia	5	Sri Lanka	1
India	0	Thailand	25
Indonesia	1		

These soils are generally sandy in texture and mostly occur on higher terraces and plateaus bearing laterites or pisolitic concretions. The majority of low TP Thai samples are from the Khorat Plateau region and are the soils derived from weathered Mesozoic sandstones. The five Cambodian samples are also derived from similar geologic formations. As the Japanese experience tells us that TP levels below 40 mg/100 g soil are critical for normal rice growth (see chapter 3), these low TP soils in tropical Asia would seem to suffer from serious phosphorus deficiency.

Available Phosphorus (AP)

(a) AP by Bray-Kurtz No. 2 method. (Bray-P) (Fig. 5.7 and Table 5.8)

As seen in the histogram the distribution of Bray-P is strongly positively skewed. More than 50 percent of the entire samples have less than 1.5 mg P_2O_5/100 g soil. The overall mean is only 3.8 mg, which is much less than the means for Japanese and Mediterranean soils.

The mean for Burmese samples is very high, a little more than 10 mg. But one of the Burmese samples has an exceptionally high Bray-P, 124.5 ng, and if this one is excluded the mean is drastically lowered to

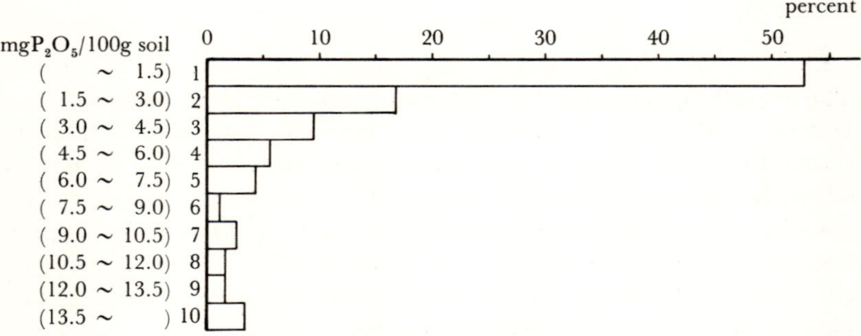

Fig. 5.7   Histogram for Available Phosphorus (Bray-P) Content of Paddy Soils in Tropical Asia

Table 5.8   Mean, Standard Deviation, Minimum and Maximum Values of Available Phosphorus (in mg P_2O_5/100 g air-dry soil)

Country	No. of Samples	Mean	Standard Deviation	Minimum	Maximum
Tropical Asia	410	3.8	10.6	tr.	124.5
Bangladesh	53	3.8	3.1	0.3	12.3
Burma	16	10.3	30.6	0.5	124.5
Cambodia	16	0.6	1.1	0.1	3.9
India	73	8.1	18.7	0.2	121.3
Indonesia	44	2.4	2.8	0.1	14.1
W. Malaysia	41	3.7	4.2	0.3	20.0
Philippines	54	3.7	4.5	0.1	17.8
Sri Lanka	33	1.4	1.3	0.1	7.4
Thailand	80	1.4	3.2	tr.	21.2
Mediter. Countries	62	7.2	6.7	tr.	31.9
Japan	84	12.9	11.7	0.1	63.0

2.7 mg. In the Indian samples, the mean is kept relatively high at 5.2 mg even after excluding two samples that have more than 100 mg Bray-P. Reasons for these high Bray-P contents are not known.

The low means for Cambodia, Thailand and Sri Lanka suggest a general paucity of available phosphorus in paddy soils. Indonesian soils are not very rich either, despite their relatively high total phosphorus content.

(b) AP by 0.2N HCl dissolution method (HCl-P) (Fig. 5.8 and Table 5.9)

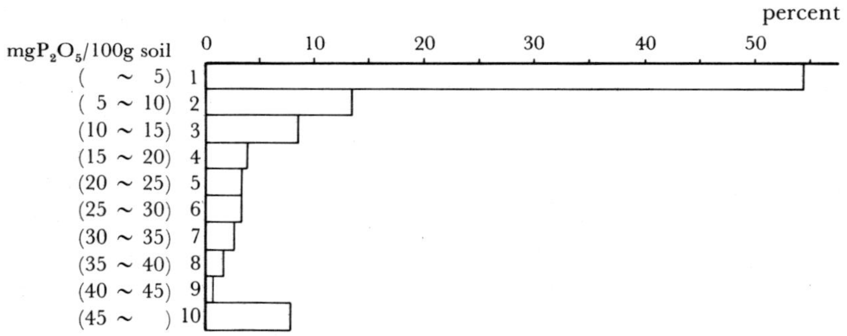

FIG. 5.8　HISTOGRAM FOR AVAILABLE PHOSPHORUS (HCl-P) CONTENT OF PADDY SOILS IN TROPICAL ASIA

TABLE 5.9　MEAN, STANDARD DEVIATION, MINIMUM AND MAXIMUM VALUES OF AVAILABLE PHOSPHORUS (HCl-P) (in mg P$_2$O$_5$/100 g air-dry soil)

COUNTRY	NO. OF SAMPLES	MEAN	STANDARD DEVIATION	MINIMUM	MAXIMUM
Tropical Asia	410	12.9	23.0	0.1	205.9
Bangladesh	53	21.0	21.3	0.1	75.4
Burma	16	24.1	47.1	0.6	190.0
Cambodia	16	1.9	5.0	0.2	20.4
India	73	21.9	32.8	0.1	205.9
Indonesia	44	10.0	13.2	0.4	51.6
W. Malaysia	41	8.2	11.9	0.2	63.0
Philippines	54	13.4	17.6	0.5	83.7
Sri Lanka	33	9.0	23.3	0.2	135.0
Thailand	80	4.7	13.7	0.1	112.6
Mediter. Countries	62	—	—	—	—
Japan	84	46.5	34.7	0.1	166.0

The level of HCl-P is always higher than that of Bray-P, because of the more drastic nature of the method. But the general parallelism between the two is obvious.

The high Burmese and Indian means are lowered to 13.1 mg and 17.6 mg, respectively, after the exclusion of the exceptional cases. Soils from Cambodia and Thailand are still critically low in this form of available phosphorus. Malaysia, Sri Lanka, and Indonesia soils are moderate, and Philippine and Bangladesh soils are good, though their HCl-P contents are still very low in comparison with Japanese soils.

Cation Exchange Capacity (CEC) (Fig. 5.9 and Table 5.10)

CEC is one of the most important fertility characteristics and, like pH, has many covarying characters, such as texture, base status, and clay composition.

The histogram shows a positive skew, and the overall mean of 18.6 me/100 g soil is far from the mode, which corresponds to 5–10 me. CEC ranges from 0.6 me for a sandy Khorat Plateau soil in Thailand to 56 me for a grumusol in India.

High means and relatively low variances are seen for the samples from Indonesia and the Philippines. The soils of both the countries are generally rich in good quality clays. Sri Lanka and Bangladesh soils have low mean CEC values. The mode of the Thai soils is 1, which corresponds to CEC values less than 5 milli-equivalent. Most soils falling into this class come from the Khorat Plateau. The maximum figures for most countries are for grumusols or grumusolic alluvial soils.

In spite of the high organic matter content, both the mean and the maximum figures for Malaysian soils are not remarkably high. The reason for this may be the peaty, not well-humified nature of the organic matter.

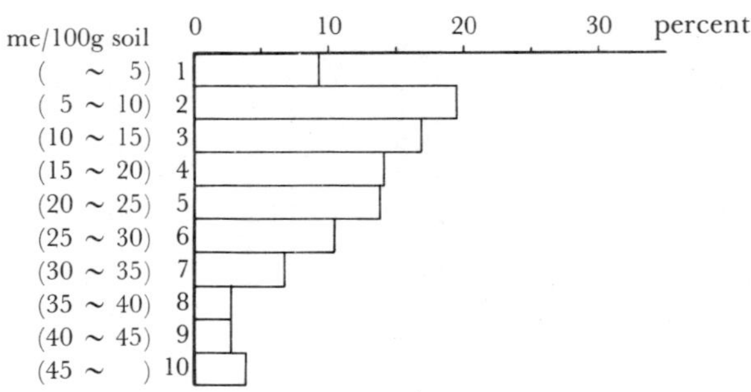

FIG. 5.9   HISTOGRAM FOR CATION EXCHANGE CAPACITY OF PADDY SOILS IN TROPICAL ASIA

TABLE 5.10   MEAN, STANDARD DEVIATION, MINIMUM AND MAXIMUM VALUES OF CATION EXCHANGE CAPACITY (in me/100 g air-dry soil)

COUNTRY	NO. OF SAMPLES	MEAN	STANDARD DEVIATION	MINIMUM	MAXIMUM
Tropical Asia	410	18.6	12.0	0.6	56.0
Bangladesh	53	12.9	7.0	3.6	32.5
Burma	16	18.6	9.1	5.6	33.4
Cambodia	16	14.6	10.4	2.1	34.8
India	73	22.0	13.7	1.7	56.0
Indonesia	44	26.1	12.6	3.0	55.4
W. Malaysia	41	15.9	8.6	2.9	33.2
Philippines	54	27.0	11.8	6.3	52.3
Sri Lanka	33	11.5	7.1	3.0	34.2
Thailand	80	14.4	10.5	0.6	54.3
Mediter. Countries	62	18.0	11.6	1.6	37.0
Japan	84	20.3	7.0	8.8	38.0

Exchangeable Calcium (Ex-Ca) (Fig. 5.10 and Table 5.11)

Exchangeable calcium is the major component of the total exchangeable cations. In our analysis part of the free calcium carbonate present in some sample soils is measured together with exchangeable calcium, and this explains why some of the maximum figures are even greater than the corresponding maximum CEC values.

Reflecting their acid nature, Malaysian soils have the lowest mean Ex-Ca. Cambodia and Sri Lanka samples also show low means. Indonesia, India, and the Philippines have higher mean values, with Philippine soils having the highest mode and the minimum variance.

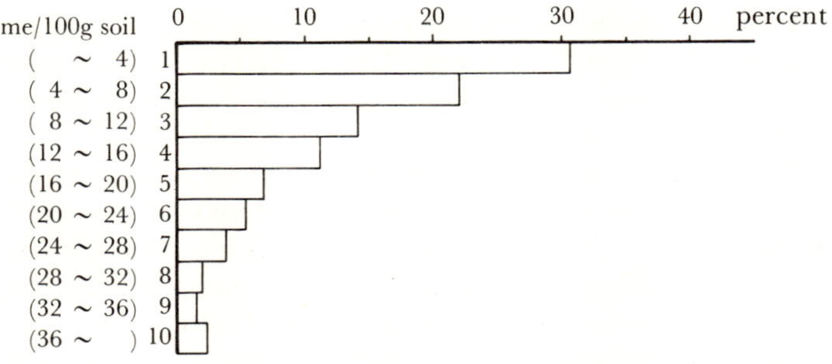

FIG. 5.10   HISTOGRAM FOR EXCHANGEABLE CALCIUM CONTENT OF PADDY SOILS IN TROPICAL ASIA

TABLE 5.11   MEAN, STANDARD DEVIATION, MINIMUM AND MAXIMUM
VALUES OF EXCHANGEABLE CALCIUM (in me/100 g air-dry soil)

COUNTRY	NO. OF SAMPLES	MEAN	STANDARD DEVIATION	MINIMUM	MAXIMUM
Tropical Asia	410	10.4	9.9	0.1	64.7
Bangladesh	53	7.8	7.8	0.2	33.1
Burma	16	10.1	10.6	1.5	38.7
Cambodia	16	5.4	3.6	0.7	12.5
India	73	15.0	10.9	0.2	38.6
Indonesia	44	17.8	14.2	1.2	64.7
W. Malaysia	41	3.9	3.4	0.1	12.3
Philippines	54	14.8	7.1	2.7	29.3
Sri Lanka	33	5.4	5.0	0.5	22.0
Thailand	80	7.2	7.5	0.2	43.8
Mediter. Countries	62	15.9	13.1	2.0	38.3
Japan	84	9.3	5.3	1.6	44.9

Among the soils containing high Ex-Ca (including those having
free carbonates), there are two broad groups. One is grumusols, and the
other is alluvial soils on recent calcareous alluvia. The latter is typically
seen in the Gangetic alluvial region in northern India and Bangladesh.
Recent alluvia of the Solo River in East Java are also calcareous.
Grumusols (or regurs) occur widely in India, in Upper Burma, and in
Central and East Java, and as patches in Thailand, Cambodia, Sri
Lanka (Dry Zone), and the Philippines.

Exchangeable Magnesium (Ex-Mg) (Fig. 5.11 and Table 5.12)

The overall mean Ex-Mg content is 5.5 me/100 g soil, which constitutes about 30 percent of the mean CEC. This computation is possible
because free Mg salts seldom occur, except for a few soils recently reclaimed from marine clays.

Soils containing relatively more Ex-Mg than Ex-Ca can be considered to have been derived from marine or brackish alluvia. Ex-Mg
content greater than 5 me and Ex-Mg/Ex-Ca ratio greater than unity
may be used for differentiating soils of marine or brackish alluvia origin.
Acid sulfate soils, however, do not satisfy either criteria.

Grumusols and related soils derived from basic igneous rocks contain a high amount of Ex-Mg, but in these soils Ex-Ca content is even
higher.

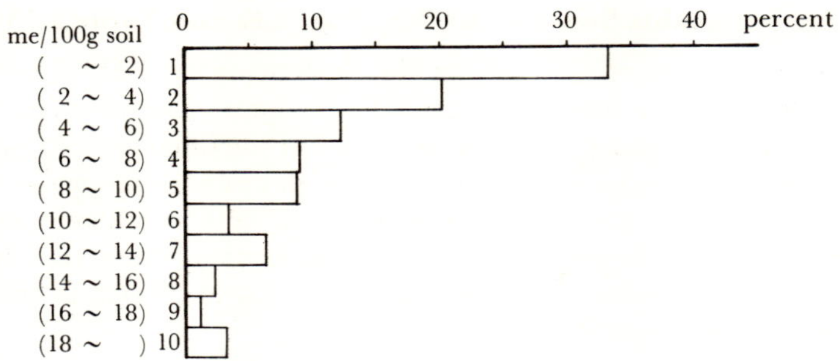

FIG. 5.11   HISTOGRAM FOR EXCHANGEABLE MAGNESIUM CONTENT OF PADDY SOILS IN TROPICAL ASIA

TABLE 5.12   MEAN, STANDARD DEVIATION, MINIMUM AND MAXIMUM VALUES OF EXCHANGEABLE MAGNESIUM (in me/100 g air-dry soil)

COUNTRY	NO. OF SAMPLES	MEAN	STANDARD DEVIATION	MINIMUM	MAXIMUM
Tropical Asia	410	5.5	5.3	0.1	34.1
Bangladesh	53	2.7	2.4	0.1	9.6
Burma	16	7.7	6.1	0.7	20.8
Cambodia	16	3.2	3.5	0.1	12.0
India	73	6.5	5.1	0.3	24.4
Indonesia	44	6.3	3.9	1.1	15.5
W. Malaysia	41	5.2	6.4	0.4	28.7
Philippines	54	9.3	6.0	1.0	29.3
Sri Lanka	33	3.5	3.5	0.1	14.7
Thailand	80	4.3	5.5	0.1	34.1
Mediter. Countries	62	4.6	4.0	0.8	15.0
Japan	84	2.8	1.8	0.5	9.4

The frequency of these two groups of soils in the samples explains the mean values for different countries. The Philippines, India, and Indonesia have many samples of grumusolic nature. In Burma, deltaic soils are of marine clay origin, and Upper Burma soils are grumusolic. Bangladesh, Cambodia, and Sri Lanka samples have only a few of these soils. The Bangkok Plain soils have pushed up slightly the mean for Thailand.

The overall maximum figure is for a soil of marine clay origin in the Bangkok Plain of Thailand, which has about 10 me of water soluble magnesium salts.

Exchangeable Sodium (Ex-Na) (Fig. 5.12 and Table 5.13)

Sodium is one of the most mobile elements in the soil. Therefore, where the climate is humid, sodium is leached out of all soils regardless of their physiographic position. This is the case for Japanese soils, whose mean Ex-Na content is only 0.4 me/100 g soil. But where climate is dry or alternately wet and dry, sodium tends to be accumulated in the lower physiographic position as a result of leaching from the higher positions. Thus, paddy soils in lowlands are apt to be rich in sodium in most parts of tropical Asia.

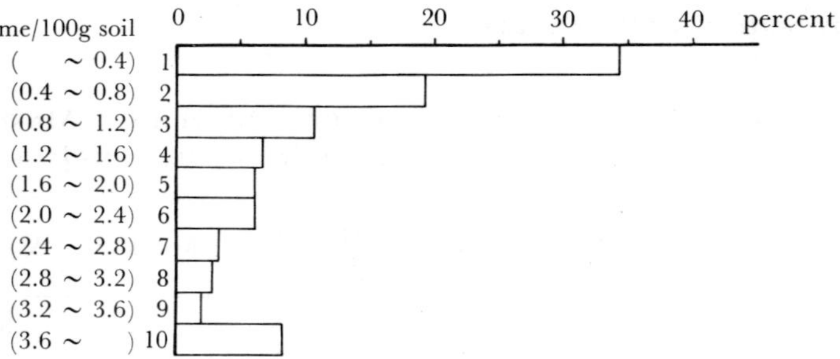

FIG. 5.12   HISTOGRAM FOR EXCHANGEABLE SODIUM CONTENT OF PADDY SOILS IN TROPICAL ASIA

TABLE 5.13   MEAN, STANDARD DEVIATION, MINIMUM AND MAXIMUM VALUES OF EXCHANGEABLE SODIUM
(in me/100 g air-dry soil)

COUNTRY	NO. OF SAMPLES	MEAN	STANDARD DEVIATION	MINIMUM	MAXIMUM
Tropical Asia	410	1.5	3.0	tr.	46.0
Bangladesh	53	0.9	1.2	0.1	5.3
Burma	16	0.6	0.8	0.1	3.5
Cambodia	16	0.3	0.3	0.1	0.9
India	73	2.4	2.0	0.1	8.3
Indonesia	44	1.5	1.5	0.2	6.3
W. Malaysia	41	1.5	4.1	0.1	26.0
Philippines	54	2.2	1.8	0.3	8.9
Sri Lanka	33	0.5	0.6	0.1	2.7
Thailand	80	1.4	5.3	tr.	46.0
Mediter. Countries	62	1.0	1.3	0.1	6.4
Japan	84	0.4	0.4	0.1	2.9

Indian soils have the highest mean Ex-Na content reflecting the climate. The high mean value for the Philippines seems partly due to the nature of parent materials. Parent material seems to have played a role in setting the lowest mean for Cambodian soils, too.

There are cases of abnormally high Ex-Na content. The maximum figures in Thailand and Malaysia are explained by the occurrence of free Na salts in the soils of recent marine clay origin, which also contain free Mg salts. Some of the soils from the Khorat Plateau region of Thailand are known to have a high sodium content. This is explained by the occurrence of a salt bed intercalated in the sandstone basement. Sodium chloride is supplied to the surface soil by the capillary rise of ground water during the dry season (see chapter 9).

When sodium is present in considerable amounts as exchangeable cations, it tends to give a high pH as a result of hydrolysis. Besides the alkaline soils in India referred to earlier, there are soils with a pH above 7 and high relative percentage of Ex-Na in spite of their slightly unsaturated exchange complex.

One should note here that most of our samples were collected during the dry season, when sodium content is at its maximum. At the onset of the rainy season, many of the sodium ions would be leached out, and the Ex-Na status during the rice-growing season could be different from our findings.

Exchangeable Potassium (Ex-K) (Fig. 5.13 and Table 5.14)

It is obvious from the table that the Ex-K level is much lower than that of Ex-Na. This is partly because of the relative stability of such K-bearing minerals as orthoclase and mica in the soil system in comparison with Na-bearing plagioclase. There may also be mechanisms operating in the soil which suppress the potassium level in the soil solution, such as potassium fixation by 2:1 clays. Selective uptake of potassium by plants must also be considered.

If we disregard the difference in the general level, however, Ex-K shows behavior similar to that of Ex-Na. Indian soils have the highest mean Ex-K and Cambodian soils the lowest. The overall maximum content of Ex-K is seen in a Thai soil derived from marine clay, which also has the highest amounts of Mg and Na.

Some soils contain only traces of Ex-K. But even in such soils rice plants can take up potassium from irrigation water, which usually contains 1 to 2 ppm or more of K. If we assume, as stated in chapter 3, that the water requirement for a single crop of rice is 1,000 mm, only 1 ppm of an element in the irrigation water would supply about 10 kg of that element per hectare per crop. This supply through irrigation water would make potassium a rare limiting factor in crop growth.

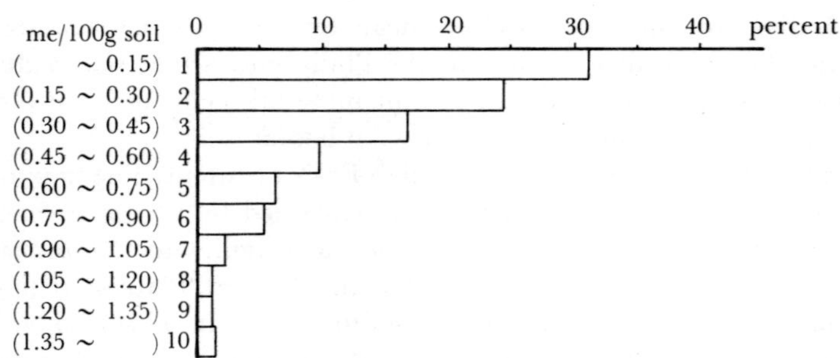

FIG. 5.13   HISTOGRAM FOR EXCHANGEABLE POTASSIUM CONTENT OF PADDY SOILS IN TROPICAL ASIA

TABLE 5.14   MEAN, STANDARD DEVIATION, MINIMUM AND MAXIMUM VALUES OF EXCHANGEABLE POTASSIUM
(in me/100 g air-dry soil)

Country	No. of Samples	Mean	Standard Deviation	Minimum	Maximum
Tropical Asia	410	0.4	0.3	tr.	2.6
Bangladesh	53	0.3	0.2	tr.	0.8
Burma	16	0.4	0.2	0.1	0.7
Cambodia	16	0.2	0.2	tr.	0.8
India	73	0.5	0.4	0.1	2.3
Indonesia	44	0.4	0.3	0.1	1.2
W. Malaysia	41	0.4	0.4	0.1	1.8
Philippines	54	0.5	0.3	0.1	1.7
Sri Lanka	33	0.2	0.2	tr.	0.7
Thailand	80	0.3	0.4	tr.	2.6
Mediter. Countries	62	0.6	0.7	0.1	2.8
Japan	84	0.4	0.3	tr.	1.4

Available Silica (Fig. 5.14 and Table 5.15)

Silica is not thought to be essential to most agricultural crops, but for rice it is regarded as essential. The requirement for silica-containing fertilizers is judged by the pH 4 acetate-soluble silica in the soil. The criterion adopted in Japan is as follows (Imaizumi and Yoshida, 1958):

Response	Available SiO_2, mg/100 g soil
Always negative	> 13.0

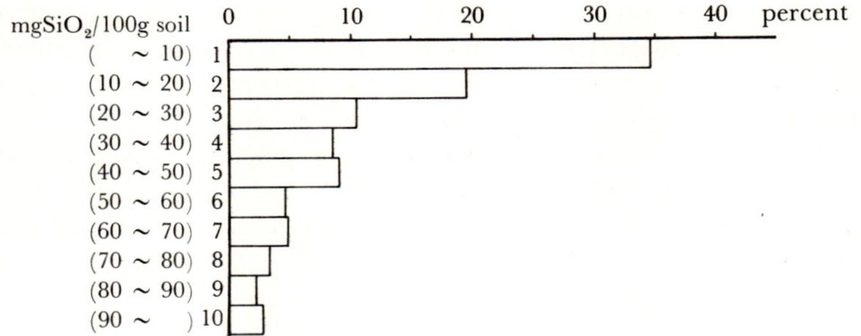

FIG. 5.14   HISTOGRAM FOR AVAILABLE SILICA CONTENT OF PADDY SOILS IN TROPICAL ASIA

TABLE 5.15   MEAN, STANDARD DEVIATION, MINIMUM AND MAXIMUM VALUES OF AVAILABLE SILICA
(in mg SiO_2/100 g air-dry soil)

COUNTRY	NO. OF SAMPLES	MEAN	STANDARD DEVIATION	MINIMUM	MAXIMUM
Tropical Asia	410	27.0	25.5	0.1	119.6
Bangladesh	53	12.9	11.4	2.2	49.0
Burma	16	22.1	15.8	4.1	54.1
Cambodia	16	11.3	10.6	1.2	39.6
India	73	34.7	24.8	0.3	87.1
Indonesia	44	62.9	25.6	5.4	119.6
W. Malaysia	41	10.4	10.0	1.0	53.5
Philippines	54	45.4	23.9	6.7	111.7
Sri Lanka	33	21.6	20.4	1.9	81.3
Thailand	80	12.2	11.1	0.1	54.0
Mediter. Countries	62	23.9	17.8	4.3	82.7
Japan	84	19.5	21.9	3.5	114.4

Positive or negative	$10.5 \sim 13.0$
Always positive	< 10.5

The same criterion may not be applicable to tropical paddy soils, because the rate of silica release from soil minerals or weathering intensity seems to be higher in the tropics. This might also explain the high overall mean of 27 mg, higher than the Japanese mean, which is already high because of the wide distribution of volcanic ejecta.

Coefficients of variation (c.v.) are lower for the countries having high mean values, for example, Indonesia and the Philippines. India and

Burma have medium means and c.v. The other countries have relatively low means and high variance. This, together with the histogram showing strong positive skewness, suggests that many samples from these latter countries would have critically low silica contents.

By setting a threshold value for available silica tentatively at 5 mg/ 100 g soil, which is about half of the Japanese value, the number of samples having available silica less than this value is listed below for each country:

Bangladesh	15	Malaysia	11
Burma	1	Philippines	0
Cambodia	6	Sri Lanka	5
India	6	Thailand	25
Indonesia	0		

These soils are, almost without exception, sandy in texture and low in pH. They occur most widely in the Khorat Plateau region of Thailand, and in the marginal plains and Pleistocene terraces (Madhupur and Barind Tracts) of Bangladesh.

The soils with very high silica contents, more than 40 mg, are either those derived from recent volcanic ejecta in the Philippines and Indonesia or those of grumusolic nature occurring in the climatic regions with a distinct dry season. The soils of swampy lands have moderate amounts of available silica.

5.2 DESCRIPTION OF MATERIAL CHARACTERS

In this section characters that are more directly connected with the soil material, that is, textural composition, clay mineralogical composition, and total chemical composition, are described.

Textural Composition

The number and description of the samples are the same as before. Again only the surface soil samples are studied, for the subsurface and subsoil materials in the rooting zone of the rice plant are, in most cases, similar to the surface soil material.

The mechanical (or textural) composition was determined either by the hydrometer or the pipette method, taking the following inter-national grain size (equivalent diameter) limits for different textural separates:

Coarse sand	2–0.2 mm
Fine sand	0.2–0.02 mm
Silt	0.02–0.002 mm
Clay	< 0.002 mm

The dispersion of clay was complete after successive hydrogen peroxide and hydrochloric acid treatments.

The textural classes presently used in Japan are defined as shown in Figure 5.15 a in terms of clay, silt, and sand (coarse and fine sands combined) (see Tommerup, 1934). For comparison, the USDA system of textural classification (Soil Survey Staff, 1951) is also shown in Figure 5.15 b, as it is widely used in the countries of tropical Asia. But in the latter system, as the silt is defined as between 0.05 and 0.002 mm in equivalent diameter, it is not possible to convert the Japanese classes to the USDA classes.

The textural data of the samples are plotted in the triangular diagram by countries, as shown in Figure 5.16 and the distribution of the samples among different textural classes is given in Table 5.16. From the figure it is clear that Sri Lanka samples are concentrated in the lower left corner, that is, they are of high sand and low silt contents. The same is evident in the table, which indicates that 72.6 percent of the Sri Lanka samples are classified as sandy (S, LS, SL, SCL, and SC), leaving only 15.2 percent and 12.1 percent in LiC and HC classes, respectively. The sandy and poor-in-silt nature of the sample soils can be explained by the fact that the greater part of paddy soils in Sri

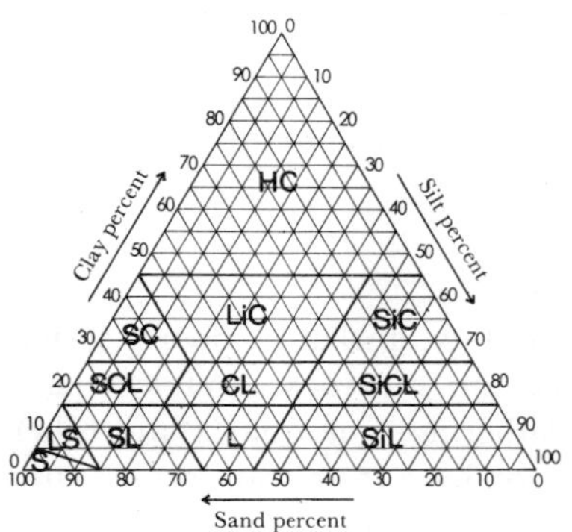

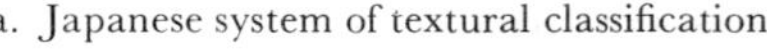
Sand percent

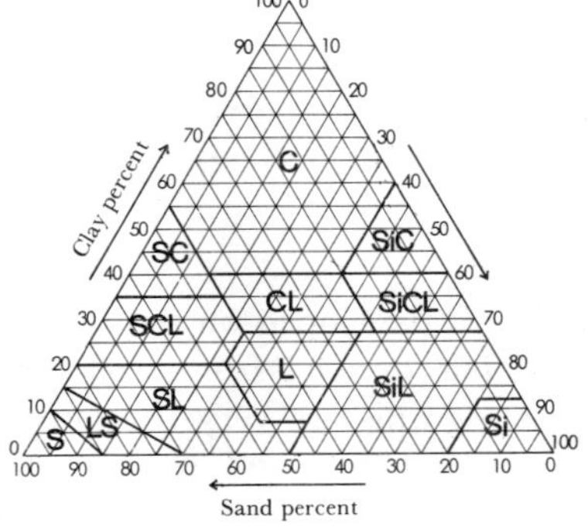

Sand percent

a. Japanese system of textural classification   b. USDA system of textural classification

Key:

S—Sand
LS—Loamy sand
SL—Sandy loam
L—Loam
SiL—Silt loam
Si—Silt
SCL—Sandy clay loam

CL—Clay loam
SiCL—Silty clay loam
SC—Sandy clay
LiC—Light clay
C—Clay
SiC—Silty clay
HC—Heavy clay

Fig. 5.15   Triangular Diagrams for Textural Classification

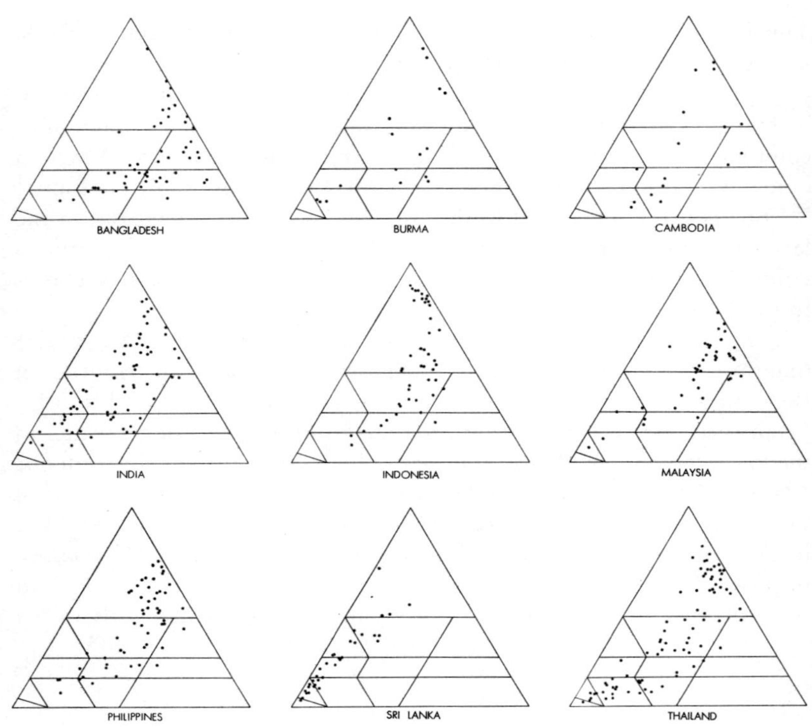

Fig. 5.16   Distribution of Sample Soils in the Triangular Diagram for Textural Classification

Lanka is developed on residual or local alluvial, that is, poorly sorted materials derived from well-weathered gneissic rocks. While coarse quartz grains are very resistant to weathering, fine quartz and other minerals of fine sand and silt size are totally lost or transformed into clay by intensive chemical weathering. In fact, many Sri Lanka soils contain less of not only silt but also fine sand than coarse sand.

In Figure 5.16 another characteristic pattern is seen in the case of Bangladesh soils, which generally are found toward the lower right side of the triangle, that is, toward the silty textural classes. Table 5.16 indicates that about 40 percent of the total samples from Bangladesh fall into the silty classes (SiL, SiCL, SiC). The silty nature of soil material is a common feature of both Ganges and Brahmaputra sediments.

The abundance of heavy-textured soils is noticeable in the case of

TABLE 5.16   PERCENTAGE DISTRIBUTION OF SAMPLE SOILS AMONG
DIFFERENT TEXTURAL CLASSES

TEXTURAL CLASS	S	LS	SL	L	SiL	SCL	CL	SiCL	SC	LiC	SiC	HC
Bangladesh	0	0	3.8	5.7	1.9	0	19.8	19.8	0	7.5	17.0	24.5
Burma	0	0	18.8	0	0	6.2	12.5	12.5	0	18.8	0	31.2
Cambodia	0	0	12.5	12.5	0	6.2	12.5	0	0	9.4	12.5	34.4
India	0	1.4	2.7	1.4	0	11.6	12.3	0	8.2	23.4	2.7	36.3
Indonesia	0	0	4.5	0	0	0	10.2	2.3	0	21.6	2.3	59.1
Malaysia	0	2.4	2.4	0	0	0	7.3	0	4.9	12.2	9.8	61.0
Philippines	0	0	5.6	3.7	0	1.8	12.0	1.8	0.9	22.2	3.7	48.1
Sri Lanka	3.0	21.2	13.6	0	0	21.2	0	0	13.6	15.2	0	12.1
Thailand	1.2	5.6	15.6	7.5	0	2.5	5.0	0	0	22.5	2.5	37.5
Tropical Asia	0.5	3.3	7.8	3.4	0.2	5.0	10.1	3.5	3.2	18.3	5.4	39.3

Indonesia, the Philippines, and Malaysia, of which the former two are in the volcanic regions of insular Southeast Asia. The basic nature of volcanic ejecta is the cause of the heavy texture, because they are easily weathered to clay, without leaving sand-size grains behind. Many of the rice areas in Malaysia are reclaimed from marine and freshwater swamps which are rich not only in organic matter, as mentioned in the preceding section, but also in clay. Thus, the heavy texture of Malaysian soils is conditioned by the sedimentary process, while that of Indonesian and Philippine soils is caused by the weathering of basic parent rocks.

Among the Thai soils, sandy light-textured soils from the Northeast or Khorat Plateau region make one cluster, as seen in the lower left corner of the triangle. Another large cluster of samples seen in the upper right corner is mainly composed of the soils from the Bangkok Plain.

The heavy-textured soils from India are mostly grumusols and grumusolic alluvial soils in the deltas of such big rivers as the Ganges, Godavari-Krishna, and Cauvery. Many of the sandy clay and sandy clay loam soils are on strongly weathered old alluvial terraces or lateritic plateaus.

Cambodian and Burmese soils are rather scattered in the triangular diagrams. No general remarks can be made on the basis of such a small number of samples.

The mean and standard deviation for clay, silt, and sand content are tabulated in Table 5.17 by country. When plotted in a triangular diagram, the points representing the mean compositions of Burma, Cambodia, India, and Thailand are clustered in the center, with those

TABLE 5.17   MEANS AND STANDARD DEVIATIONS FOR SAND, SILT, AND CLAY CONTENTS (IN PERCENT)

COUNTRY	NO. OF SAMPLES	SAND		SILT		CLAY	
		Mean	S.D.	Mean	S.D.	Mean	S.D.
Bangladesh	53	25.6	20.0	42.8	13.1	31.6	17.2
Burma	16	38.2	28.4	26.0	14.2	35.7	25.2
Cambodia	16	33.9	25.5	31.4	12.7	34.7	24.3
India	73	36.4	22.8	24.3	11.2	39.3	19.6
Indonesia	44	22.5	18.3	26.3	10.8	51.2	24.2
Malaysia	41	22.5	22.0	31.3	10.6	46.1	16.1
Philippines	54	27.1	20.8	30.8	9.2	42.1	18.8
Sri Lanka	33	68.3	20.2	7.6	6.0	24.1	15.8
Thailand	80	38.2	29.3	25.2	11.1	36.7	24.6
Tropical Asia	410	33.9	26.0	27.7	13.7	38.4	21.6
Japan	155	49.2	18.2	29.6	10.6	21.2	10.1

of Indonesia, Malaysia, and the Philippines are toward the upper right, that of Bangladesh slightly toward the lower right, and the one for Sri Lanka far down on the lower left, making these stated features still clearer.

For comparison, the mean textural composition for 155 Japanese paddy soils sampled from all over the country is also given. In Japanese soils the silt content is nearly equal, the clay content is much lower, and the sand content much higher than in tropical Asian paddy soils. Low standard deviations for the Japanese soils imply homogeneity in texture.

Clay Mineral Composition

Clay mineralogical composition was semiquantitatively assessed by measuring the areas of 7 Å, 10 Å and 14 (to 15) Å peaks on the X-ray diffraction diagram for the oriented specimen of Ca-saturated and air-dried clay and calculating the area of each peak as the percentage of the total area of the three peaks. As chlorite seldom occurs, 7 Å minerals may be regarded as 1:1 type clay minerals of kaolin group, and 10 Å minerals as illite or clay micas. The 14 Å minerals consist mainly of montmorillonite, vermiculite, and Al-interlayered vermiculite-chlorite intergrades, of which montmorillonite is the most common. It tends to predominate particularly in soils containing a high amount of 14 Å minerals. Amorphous clays were not estimated. One sample from Indonesia and three samples from the Philippines did not contain crystalline clay minerals. The allophanic nature of the clay fraction of these soils has been previously established and reported (Kitagawa *et al.*, 1973).

The clay mineral composition is plotted in a triangular diagram in terms of 7, 10, and 14 Å minerals. In this diagram 10 classes are set up, as shown in Figure 5.17, based on relative abundance of the respective mineral species. Figure 5.18 shows the sample data plotted in the diagram for each country, and Table 5.18 gives the number of samples falling into each class.

As evident from the figures and the table, no sample contains 10 Å minerals in amounts greater than 75 percent of the total crystalline clays. Even samples containing 10 Å minerals as the dominant clay species, together with either 7 Å (10–7) or 14 Å minerals (10–14), are very rare.

There seem to be roughly three types of clay mineral composition:

"7-dominant" type, having either 7, 7–10, or 7–14 combinations—Cambodia, Malaysia, Sri Lanka, and Thailand,
"14-dominant" type, having either 14, 14–7, or 14–10 combinations—Indonesia and the Philippines,
"7–10–14-even" type, having more than 25 percent of all 3 clay species—Bangladesh, Burma, and India.

Of the three types, the "7-dominant" type may be regarded as the most

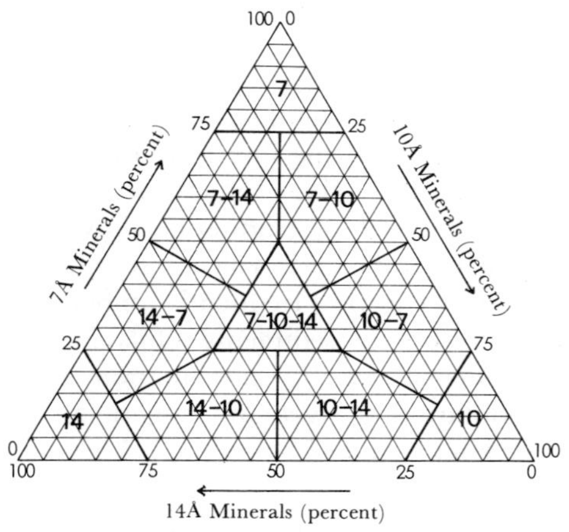

FIG. 5.17   TRIANGULAR DIAGRAM FOR CLAY MINERAL COMPOSITION: 7-7 Å minerals, 10-10 Å minerals, 14-14 Å minerals

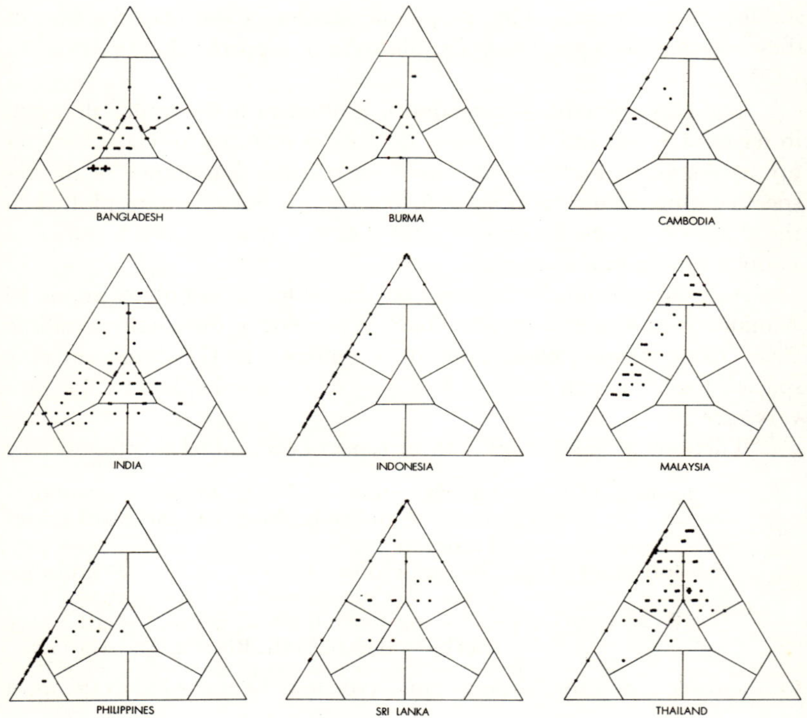

Fig. 5.18   Distribution of Sample Soils in the Triangular Diagram for Clay Mineral Composition

highly weathered and the "14-dominant" type as the least weathered, the "7–10–14-even" type as being intermediate.

Samples containing more than 75 percent 14 Å minerals ("14" in Fig. 5.17) are either grumusols or grumusolic alluvial soils derived from calcareous sediments or from basic volcanic ejecta. The "14–7" and "7–14" combinations occur most frequently among the soils developed on recent alluvial and deltaic sediments. Only Bangladesh soils have an appreciable number of "14–10" combination, and the soils having this clay composition are exclusively of Gangetic alluvia origin.

Soils containing more than 75 percent 7 Å minerals are most frequently encountered in the Wet Zone of Sri Lanka, East Coast of Malaysia, and Southern Peninsular region of Thailand. The Khorat Plateau region of Thailand and Cambodia also have many samples with this clay composition. Besides these regions, soils containing a

TABLE 5.18 PERCENTAGE DISTRIBUTION OF SAMPLE SOILS AMONG DIFFERENT CLAY MINERALOGICAL GROUPS

CLAY MINERAL GROUP	7	7-10	7-14	10	10-7	10-14	14	14-7	14-10	7-10-14
Bangladesh	0	13.2	13.2	0	6.6	0	0	10.4	26.4	30.2
Burma	0	15.6	12.5	0	2.1	2.1	0	25.0	6.2	36.4
Cambodia	31.2	0	34.4	0	0	0	0	28.1	0	6.2
India	3.4	6.2	13.7	0	9.6	4.1	8.9	22.6	8.9	22.6
Indonesia	20.4	0	18.2	0	0	0	28.4	30.7	0	0
Malaysia	34.1	2.4	24.4	0	0	0	0	39.0	0	0
Philippines	3.7	0	9.2	0	0	0	37.0	42.6	0	1.8
Sri Lanka	39.4	16.7	18.2	0	0	0	6.1	16.7	0	3.0
Thailand	18.8	21.0	46.0	0	0.6	0	2.5	6.9	0	4.2
Tropical Asia	14.8	9.1	22.0	0	2.8	0.8	10.5	22.9	5.2	10.9

large amount of kaolin minerals are commonly found on highly weathered middle and high terrace sediments, which often contain pisolitic concretions and/or lateritic nodules.

The Indonesian samples that contain more than 75 percent 7 Å minerals in their clay fraction are almost exclusively latosols. They are derived from basic pyroclastic materials and seem to contain, besides kaolin minerals, high amounts of allophane and oxide minerals (goethite and gibbsite) (Kitagawa *et al.*, 1973).

In Table 5.19 the mean and the standard deviation of content of the three clay mineral species are tabulated. When the means are plotted on a triangular diagram, two fairly tight clusters, one representing the "7-dominant" type and the other the "7–10–14-even" type, can be clearly distinguished. The "14-dominant" type does not form a tight cluster as such, but its separation from the other two clusters is clear.

Of the standard deviations, the ones for 7 Å mineral content of Thai soils and for 10 Å mineral content of Bangladesh soils are remarkably low, relative to their high means. Thus, a high kaolin content and a high illite content can be regarded as characteristic mineralogical features of Thai soils and Bangladesh soils, respectively. Very low illite (10 Å mineral) contents for Cambodia, Indonesia, and Philippine soils are also worthy of note.

Total Chemical Composition

Total content of the nine major elements (Si, Fe, Al, Ca, Mg, Mn, Ti, K, P) was determined by X-ray fluorescence spectrography using

TABLE 5.19   MEANS AND STANDARD DEVIATIONS FOR 7Å, 10Å,
AND 14Å MINERAL CONTENTS (IN PERCENT)

COUNTRY	NO. OF SAMPLES	7 Å MINERAL		10 Å MINERAL		14 Å MINERAL	
		Mean	S.D.	Mean	S.D.	Mean	S.D.
Bangladesh	53	34.3	10.7	29.2	14.4	36.5	13.6
Burma	16	38.8	13.1	26.6	7.9	34.7	12.4
Cambodia	16	60.6	19.2	4.1	8.5	35.3	17.8
India	73	34.7	16.4	23.7	8.2	41.5	20.6
Indonesia	44	46.6	29.7	0.7	15.3	52.7	30.5
Malaysia	41	59.4	22.2	9.0	2.6	31.6	23.4
Philippines	54	31.2	19.4	2.6	4.5	66.2	24.3
Sri Lanka	33	64.2	25.6	9.1	11.5	26.7	23.6
Thailand	80	59.2	16.5	12.9	12.0	27.9	15.5
Tropical Asia	410	46.4	23.3	13.9	14.4	39.7	23.8

the method proposed by Norrish and Chappel (1967). The standard
procedure finally adopted was as follows:

(1) Ignition of air-dried fine soil (<2 mm) to remove organic
matter and to destroy carbonates
(2) Fusion of the sample with flux ($Li_2B_4O_7$–Li_2CO_3–La_2O_3 mixture) to prepare the sample glass
(3) Spectrographic analyses of nine elements (Si, Fe, Al, Ca, Mg,
Mn, Ti, K, P)
(4) Calibration and calculation with a computer

Because the total of the contents of the nine oxides usually fell in the
range 97–102 percent, the contents were recalculated to make the sum
100 percent, neglecting other minor elements.

Reproducibility of the method is shown in Table 5.20, the data
for which were obtained at different times by different analysts. A
comparison between spectrographic and chemical methods is given in
Table 5.21. The results appeared to be satisfactory for classification
purposes. The rate of analysis, five to ten samples per day per person
(on an average), was also adequate for routine use.

The mean total chemical composition of the samples from the
respective countries are shown in Table 5.22. It is clear from the table
that soils from Indonesia and the Philippines have similar chemical
composition, that is, low silica content and high iron oxide, alumina,
and manganese oxide contents. They are generally rich in alkaline
earth bases and phosphorus but poor in potash. In short, they have all
the characteristics of materials weathered from basic igneous rocks.

TABLE 5.20   REPRODUCIBILITY OF SPECTROGRAPHIC METHOD FOR
TOTAL CHEMICAL ANALYSIS OF A JAPANESE PADDY SOIL

ELEMENTAL OXIDE	1	2	3
SiO_2	76.30	76.50	76.47
Fe_2O_3	3.12	2.97	3.18
Al_2O_3	15.71	15.47	15.10
CaO	0.46	0.67	0.69
MgO	0.87	0.86	1.14
MnO_2	0.10	0.12	0.13
TiO_2	0.96	0.88	0.86
K_2O	2.09	2.14	2.05
P_2O_5	0.39	0.40	0.38

TABLE 5.21   COMPARISON BETWEEN CHEMICAL AND SPECTROGRAPHIC
METHODS (a—chemical, b—spectrographic)

ELEMENTAL OXIDE	SOIL-1 (subsurface horizon of a red soil)		SOIL-2 (subsoil of the same soil)	
	a	b	a	b
SiO_2	65.99	67.26	54.03	53.23
Fe_2O_3	8.81	7.24	14.37	12.38
Al_2O_3	18.26	19.86	26.21	30.32
CaO	0.81	0.52	0.32	0.00
MgO	1.70	1.05	1.10	0.66
MnO_2	0.52	0.24	0.09	0.05
TiO_2	1.21	1.49	1.15	1.54
K_2O	1.86	1.91	1.65	1.73
P_2O_5	0.13	0.16	0.04	0.06

Soils from Cambodia and Thailand seem to have many characteristics in common; silica content is very high, while the values for iron, aluminum, manganese, alkaline earths, and phosphorus contents are among the lowest. They seem to be at the opposite end of the scale from the soils of Indonesia and the Philippines. Only in potash content is there an appreciable difference between Cambodia and Thailand. The higher potash content of Thai soils may be ascribed to the presence of deltaic soils, which have higher amounts of clay and potash-bearing 10 Å minerals and have also been affected by marine and brackish environments.

Malaysian soils are as poor in bases and manganese oxide as the

Paddy Soils in Tropical Asia

TABLE 5.22   MEAN TOTAL CHEMICAL COMPOSITION

(IN PERCENT)

COUNTRY	No. of Samples	SiO_2	Fe_2O_3	Al_2O_3	CaO	MgO	MnO_2	TiO_2	K_2O	P_2O_5
Bangladesh	53	70.85	5.74	16.92	1.01	1.18	0.08	1.01	3.09	0.12
Burma	16	69.48	5.68	18.26	1.57	0.92	0.08	0.87	3.01	0.13
Cambodia	16	81.13	4.61	11.52	0.27	0.40	0.09	1.21	0.68	0.09
India	73	71.25	6.94	14.58	1.91	1.20	0.13	1.32	2.55	0.12
Indonesia	44	60.10	10.34	22.86	3.00	1.22	0.25	1.33	0.72	0.18
Malaysia	41	74.22	2.97	18.94	0.14	0.51	0.03	1.03	2.02	0.14
Philippines	54	66.19	7.73	19.57	2.83	1.26	0.23	1.03	0.99	0.17
Sri Lanka	33	76.45	5.35	12.98	1.00	0.66	0.10	1.50	1.82	0.14
Thailand	80	80.41	3.60	12.46	0.43	0.52	0.08	0.99	1.42	0.09
Tropical Asia	410	72.16 (11.51)	5.94 (3.73)	16.34 (6.98)	1.42 (1.96)	0.92 (0.76)	0.12 (0.12)	1.14 (0.60)	1.83 (1.27)	0.13 (0.08)
Japan	155	71.77 (4.89)	6.00 (2.82)	16.26 (2.04)	1.86 (1.16)	0.75 (0.46)	0.13 (0.06)	0.88 (0.34)	2.14 (0.74)	0.22 (0.10)

Figures in parentheses are standard deviations.

soils from Cambodia and Thailand. The low pH of the soils from these countries is definitely related to the low total base contents. The main difference between Malaysian soils and Cambodian and Thai soils is that the former has lower silica and higher alumina contents. This reflects the clayey texture of Malaysian soils. Another peculiar feature of Malaysian soils is the extraordinarily low iron content in spite of the high clay content. This may be owing to the granitic nature of the parent materials. The high potash content may also be due to this granitic nature. The relatively high phosphorus content of Malaysian soils is because of the common occurrence of swampy soils, which often accumulate phosphorus by certain biological mechanisms.

Sri Lanka soils are also highly siliceous and low in iron oxide and alumina, reflecting their sandy texture. Alkaline earth and manganese contents are moderate, probably because of the less depleted Dry Zone soils. Relatively high potash, phosphorus oxide, and titanium oxide contents are related to the residual nature of the parent material weathered from gneissic rocks.

Soils from Bangladesh, Burma, and India seem to make up another group. Contents of all the elements, except potash, are medium and deviate by narrow margins from the overall means. The only peculiar feature is their high potash content, which coincides, well with the particularly high 10 Å mineral contents of these three countries (24–29 percent).

If we compare the overall mean composition for the tropical Asian paddy soils with the overall mean composition for the 155 Japanese paddy soils previously cited the similarity in the content of the three macro-elements—SiO_2, Fe_2O_3, Al_2O_3—is very striking. However, difference between the two groups can be noted in the figures for standard deviation. Japanese soils are much more homogeneous, and, therefore, the standard deviation is much smaller. This is because latosols or such strongly weathered soils as found in the Khorat Plateau of Thailand are absent in Japan.

Of the other elements, there is a slight difference in the content of CaO, TiO_2, and P_2O_5 between the two groups. Though these differences are not statistically significant, they are worthy of comment. In spite of stronger acidity, the mean CaO content is higher for Japanese soils. The greater part of such CaO must be present in unweathered primary minerals. The higher TiO_2 content in tropical Asian soils is a reflection of the more intensive weathering, which causes residual accumulation of such titanium-bearing minerals resistant to weathering as rutile and anatase. The difference in phosphorus content is very noticeable. As mentioned earlier, phosphorus is so readily fixed by iron and aluminum that its content in the soil is modified rather easily by fertilizer application. The higher phosphorus content in Japanese soils, thus, most probably is due to the practice of applying fertilizer.

A brief note for each element follows, including a ten-class histogram prepared as in the preceding section.

(a) SiO_2 (Fig. 5.19)

The histogram shows the mode at class 4 or silica content of 65–70 percent, and a small peak at class 10 corresponding to silica content of 95–100 percent. The highly siliceous samples falling into classes 9 and 10 (SiO_2 content > 90 percent), are as follows:

Bangladesh	2	Malaysia	2
Burma	0	Philippines	0
Cambodia	6	Sri Lanka	4
India	2	Thailand	25
Indonesia	1		
		Total	42

Most of the highly siliceous samples come from either Thailand or Cambodia. They are derived from similar geologic formations (Mesozoic sandstones) widely distributed from the Khorat Plateau of Thailand to Cambodia. Highly weathered soils developed on laterite-bearing high terraces and plateaus are also sandy and siliceous and are included in the above figures.

Soils containing less than 55 percent SiO_2 (class 1) are mostly latosols from Indonesia, which have high iron oxide and alumina content and low alkaline earth content. But a few soils from Indonesia that have developed on unweathered volcanic sands also fall into this class. Thus the class 1 samples consist of two very contrasting groups of soils,

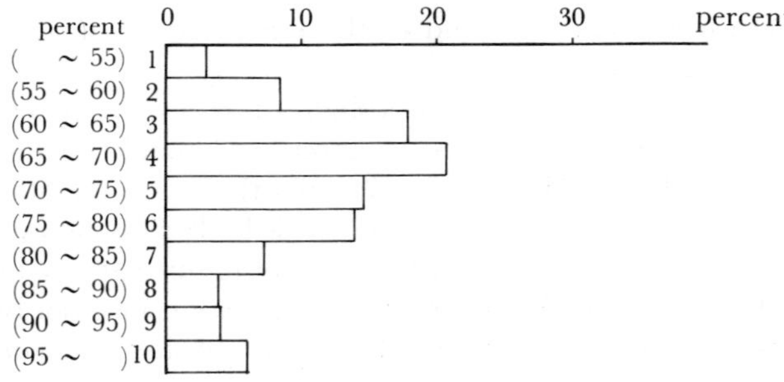

FIG. 5.19   HISTOGRAM FOR TOTAL SILICA OF PADDY SOILS IN TROPICAL ASIA

one, the majority group, of strongly weathered clayey soils of latosol type and the other of relatively unweathered soils of high potential.

(b) Fe_2O_3 (Fig. 5.20)

The least siliceous latosolic soils fall into the most iron-rich classes of the histogram (classes 8, 9, and 10), while the most siliceous ones fall into class 1. Thus, silica and iron oxide content are highly negatively correlated.

Malaysian soils have a very low mean iron content. Their distribution in the histogram is as follows: 9 in class 1; 22 in class 2; 7 in class 3; and 3 in class 4. None of the samples contain more than 8 percent of Fe_2O_3.

(c) Al_2O_3 (Fig. 5.21)

Unlike the other elements, the histogram for alumina is negatively skewed, having its mode at class 7, which corresponds to Al_2O_3 content of 18–21 percent, and its overall mean at 16.3 percent. Alumina and silica contents are negatively correlated. The pattern of the histogram for alumina is a reversal of that for silica. The small peak at class 1 is mostly because of the very highly siliceous samples containing >95 percent silica, and the samples falling into class 10 are the same latosols referred to under silica.

(d) CaO (Fig. 5.22)

The histogram is strongly positively skewed. Forty percent of the total samples fall into class 1, that is, they have CaO contents of less than 0.5 percent. As stated earlier, Cambodian, Malaysian, and Thai samples are poorest in alkaline earth elements. The distribution of the samples from these three countries among the different classes of the histogram is as follows:

Class	1	2	3	4–10
Cambodia	13	2	1	0
Malaysia	40	1	0	0
Thailand	56	22	0	2

The number of samples from each country falling into class 10 (CaO content >4.5 percent) are as follows: two from Burma, four from India, nine from Indonesia, six from the Philippines, and one from Sri Lanka. They are soils derived either from relatively unweathered basic volcanic sands (Indonesia, Philippines, and Sri Lanka) or from calcareous sediments (Burma and India), some of which are grumusolic.

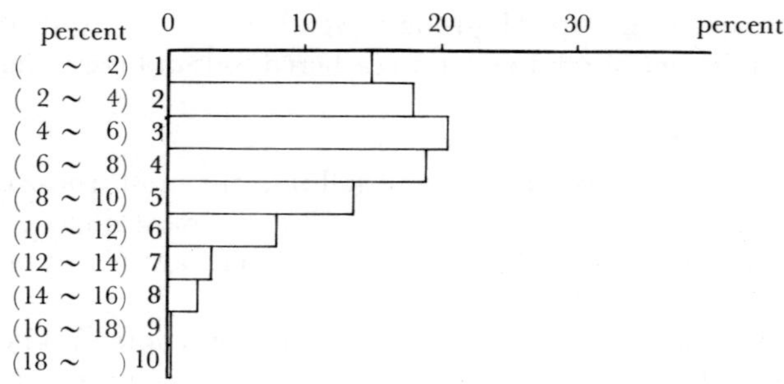

FIG. 5.20   HISTOGRAM FOR TOTAL IRON OXIDE OF PADDY SOILS IN TROPICAL ASIA

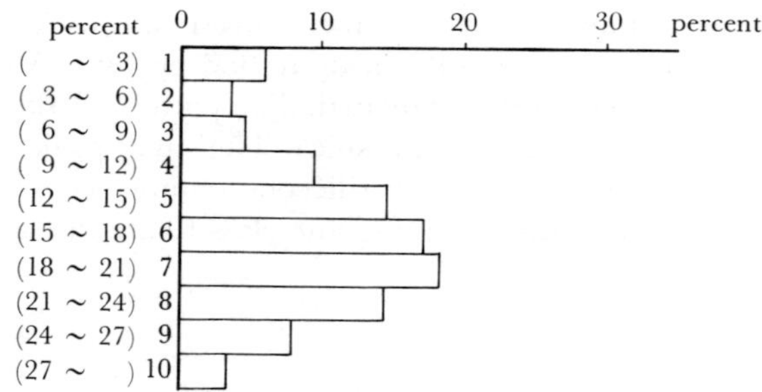

FIG. 5.21   HISTOGRAM FOR TOTAL ALUMINA OF PADDY SOILS IN TROPICAL ASIA

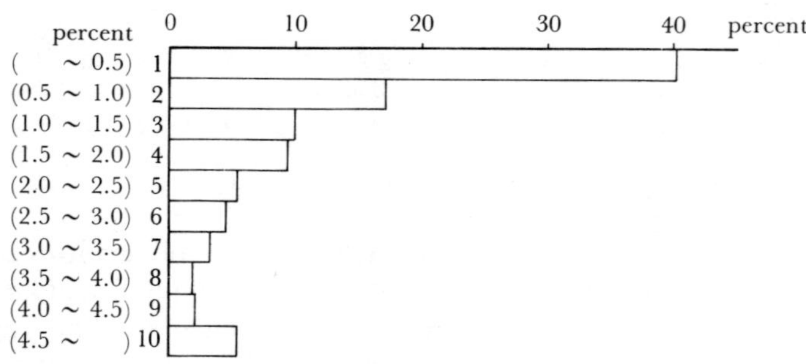

FIG. 5.22   HISTOGRAM FOR TOTAL CALCIUM OXIDE OF PADDY SOILS IN TROPICAL ASIA

(e) MgO (Fig. 5.23)

Generally, magnesium behaves similarly to calcium, both being the constituents of basic primary minerals. There is, however, an important difference in that magnesium is present in such 14 Å clay minerals as montmorillonite and vermiculite as one of the constituent elements while calcium is not. Thus, soils containing an appreciable amount of 14 Å minerals usually have 1 percent or more magnesium. But in the case of soils with 7 Å minerals as the dominant clay species, there is no mechanism to retain magnesium, and its content drops to a very low level. The minimum at class 2 of the histogram may be because of this effect.

Very high MgO content (say > 2 percent) is found in soils containing high amounts of CaO or in those recently reclaimed from marine clays.

(f) MnO₂ (Fig. 5.24)

Manganese is a mobile element especially in paddy soils, as it is easily reduced and acquires high solubility. The general resemblance of the histograms for MnO_2 and CaO may be related to the similarity in their behavior in paddy soils.

High manganese content is seen for the soils derived from basic volcanic ejecta in the Philippines and Indonesia. Manganese content of Malaysian soils is exceptionally low, with 37 of 41 sample soils falling into class 1; and the maximum content is only 0.11 percent, which is lower than the overall mean for the samples from tropical Asia as a whole. As is the case with iron, the paucity of the element in the granitic parent material is probably the primary cause of this.

(g) TiO₂ (Fig. 5.25)

Titanium is one of the most immobile elements found in the soil. It usually occurs as independent minerals, like rutile and anatase, which are extremely resistant to weathering. Therefore, its content in the soil is primarily governed by that in the parent material.

The histogram shows a very high concentration of samples at class 4, which corresponds to a TiO_2 content of 0.9–1.2 percent. One Thai soil and one Indonesian soil show exceptionally high TiO_2 content, 7.3 and 6.1 percent, respectively. It is difficult to find any explanation for this.

(h) K₂O (Fig. 5.26)

Potassium is found not only in such primary minerals as micas and orthoclase feldspars, but also, like magnesium, in secondary clay min-

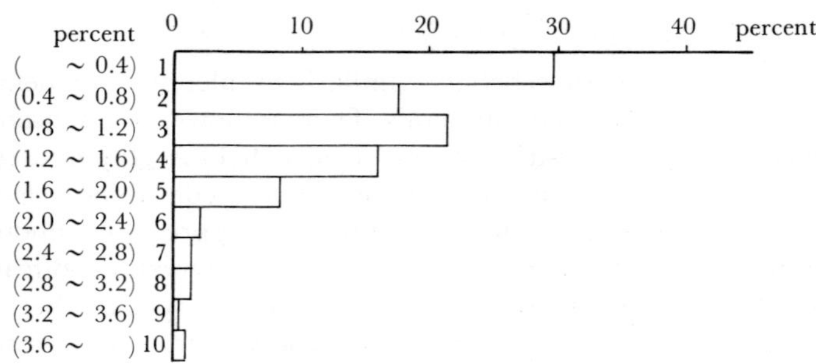

FIG. 5.23   HISTOGRAM FOR TOTAL MAGNESIUM OXIDE OF PADDY SOILS IN TROPICAL ASIA

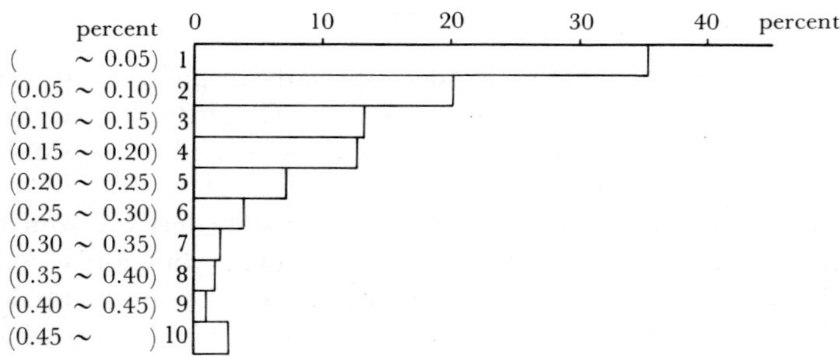

FIG. 5.24   HISTOGRAM FOR TOTAL MANGANESE OXIDE OF PADDY SOILS IN TROPICAL ASIA

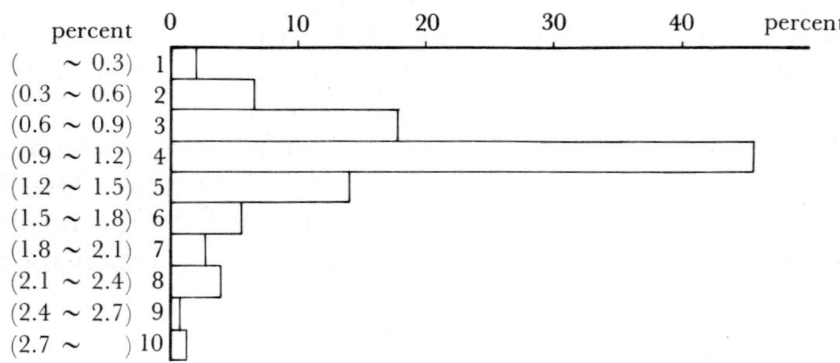

FIG. 5.25   HISTOGRAM FOR TOTAL TITANIUM OXIDE OF PADDY SOILS IN TROPICAL ASIA

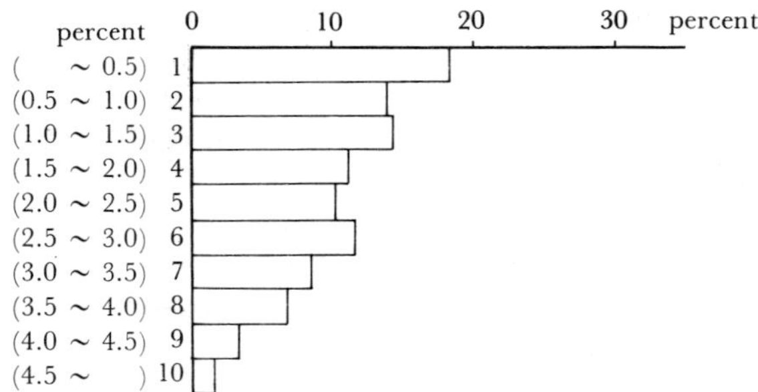

FIG. 5.26   HISTOGRAM FOR TOTAL POTASSIUM OXIDE OF PADDY SOILS
IN TROPICAL ASIA

erals. The 10 Å minerals contain about 10 percent potash in their
interlayer spaces. As these potash-containing minerals are relatively
resistant to weathering, soils with a very low potash content are not
common, even among strongly leached soils. Thus, the histogram shows
a more uniform distribution of the samples among the different classes
than is the case for other elements.

Very high potash content occurs either in soils having a high
amount of illite (for example, Bangladesh soils), or in relatively unweathered sandy soils derived from granitic or gneissic rocks (for example, Malaysian and Sri Lankan soils). Samples poor in potash are
most common among strongly weathered sandy soils from Cambodia
and the Khorat Plateau region of Thailand, and also among soils derived
from basic volcanic ejecta, the potash content of which is originally low.

(i) P_2O_5 (Fig. 5.27)

The total phosphorus content determined by the X-ray fluorescence

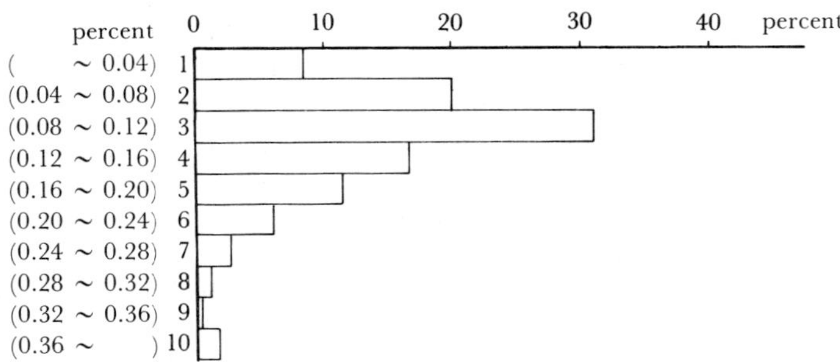

FIG. 5.27   HISTOGRAM FOR TOTAL PHOSPHORUS OXIDE OF PADDY
SOILS IN TROPICAL ASIA

method is higher than that determined by the chemical digestion method reported in the preceding section. The reason for this may be due to incomplete dissolution of phosphorus in the chemical method, and to some extent to the recalculation of the element contents by neglecting Na and other minor elements.

The general pattern of the histogram, however, is similar to the previous one, with the mode at class 3 and a small maximum at class 10. Many of the samples falling into class 10 are soils on unweathered volcanic sands from the Philippines.

Some latosols and an ando soil from Indonesia also belong to class 10. Soils very poor in phosphorus are concentrated in the Khorat Plateau region of Thailand.

5.3 CORRELATION AND REGRESSION ANALYSES OF THE SOIL DATA

In the preceding sections of this chapter a number of fertility and material characters of paddy soils in tropical Asia were described individually. In the discussions significant correlations were found between many of these characters. This shows that descriptions by each individual character alone are inadequate. Therefore, in this section correlation and regression analyses of the previously described data are presented as a preliminary step toward material classification and fertility evaluation of the sample soils.

Data and Methods Used

The data obtained from routine chemical analysis, mechanical analysis, clay mineralogical analysis, and total chemical analysis conducted on plow layer samples of 410 tropical Asian paddy soils were used in this study.

The Pearson's product-moment correlation coefficients between all pairs of these variables (or characters) were computed. In this computation, observed absolute values of the correlation coefficient larger than 0.127 were considered significant at $\alpha = 0.01$.

Multiple regression of cation exchange capacity (CEC) and some other variables upon their relevant characters were studied by assuming a linear regression model. The resulting multiple-correlation coefficients were high enough only in the case of CEC.

For assessing the contribution of climate, relief, and texture to accumulation of organic matter in the soil, Hayashi's theory of quantification No. 1 (Hayashi *et al.*, 1970) was adopted. The method aims at estimating a criterion variable that is a continuous quantitative variable, such as total carbon content in the soil, from p items of qualitative variables, such as climate and relief, each of which has $k_j(j = 1, 2 \ldots, p)$ categories. The structure of the data is shown in Table 5.23, in which the checkmark indicates the response of a sample to that particular category under each variable. Suppose that each of the categories c_{jk} is quantified as x_{jk}, and $\delta_i(jk)$ is defined as follows:

$$\begin{cases} \delta_i(jk) = 1 \text{ when } i\text{th sample shows response to } k\text{th category of } j\text{th} \\ \qquad \text{item,} \\ \delta_i(jk) = 0 \text{ otherwise.} \end{cases}$$

The score to be given to the ith sample α_i, as an estimate of the measured value A_i, is computed by the addition of the relevant x_{jk}, as follows:

$$\alpha_i = \sum_{j=1}^{p} \sum_{k=1}^{kj} \delta_i(jk) \cdot x_{jk}$$

If the x_{jk} values are given to c_{jk} so as to maximize the correlation coefficient between A and α, the α_i, computed by this equation, may be regarded as the best estimate of A_i from a set of p qualitative variables that are assigned to the ith sample.

Correlation Analysis

The correlation coefficients between all pairs of twenty-nine variables (or characters) are given in Table 5.24, in a matrix form. The variables may be grouped into six, that is, those related to base status, mechanical composition, clay mineralogical composition, organic matter status, phosphorus status, and total chemical composition.

TABLE 5.23　SCHEMATIC DATA MATRIX USED FOR HAYASHI'S THEORY OF QUANTIFICATION NO. 1

CRITERION \ ITEM (CATEGORY)	1 $c_{11}\,c_{12}\ldots\ldots c_{1k_2}$	2 $c_{21}\ldots\ldots c_{2k_2}$	....	P $c_{p1}\ldots\ldots c_{pk_p}$
A_1	√	√		√
A_2	√	√		√
A_i	√	√		√
A_n	√	√		√

TABLE 5.24 CORRELATION COEFFICIENT MATRIX BETWEEN ALL PAIRS OF TWENTY-NINE VARIABLES, WITH DECIMAL POINTS OMITTED

	pH	Ex-Ca	Ex-Mg	Ex-(Ca + Mg)	Ex-Na	Ex-K	CEC	Avail. SiO$_2$	Sand	Silt	Clay	7 A Min.	10 A Min.	14 A Min.	TC	TN	NH$_4$-N	TP	Bray-P	HCl-P	TSIO	TFEO	TALO	TCAO	TMGO	TMNO	TTIO	TKAO	TPHO
pH																													
Ex-Ca	622																												
Ex-Mg	289	534																											
Ex-(Ca + Mg)	569	943	786																										
Ex-Na	298	275	579	430																									
Ex-K	324	448	633	577	609																								
CEC	399	855	781	934	371	595																							
Avail. SiO$_2$	625	701	527	721	293	514	718																						
Sand	−016	−463	−524	−545	−243	−456	−662	−351																					
Silt	−041	−027	016	−014	−032	015	014	−074	−564																				
Clay	045	577	623	668	313	542	791	470	−854	052																			
7 A Min.	−537	−564	−474	−600	−236	−378	556	−412	283	−209	−211																		
10 A Min.	093	−125	−164	−156	−023	064	−221	−302	−066	280	−096	−230																	
14 A Min.	462	639	585	699	253	330	688	527	−282	053	308	−759	−325																
TC	−254	002	152	062	044	139	233	070	−316	110	313	−008	−159	040															
TN	−264	−057	085	−008	022	122	154	026	−297	166	255	−018	−105	−012	958														
NH$_4$-N	−117	009	084	040	021	115	143	222	−190	108	162	008	−305	100	514	586													
TP	245	194	155	203	148	320	244	415	−198	135	154	−201	−023	080	292	333	323												
Bray-P	268	079	026	068	145	350	039	154	078	−059	−057	−144	166	038	010	009	−047	333											
HCl-P	379	154	057	136	257	356	047	159	094	−001	−113	−290	253	114	−007	002	−059	436	806										
TSIO	−322	−512	−502	−573	−218	−456	−631	−603	655	−223	−651	365	002	−326	−212	−209	−243	−509	−020	−119									
TFEO	389	557	504	607	202	384	638	685	−512	100	555	−304	−133	373	045	013	114	481	−008	066	−841								
TALO	066	351	436	429	162	404	544	457	−716	244	712	−198	−030	183	330	330	321	432	−032	010	−919	657							
TCAO	553	418	157	368	109	174	278	418	−010	018	000	−449	−079	402	−063	−050	092	243	075	210	−406	289	159						
TMGO	494	434	415	482	328	379	434	414	−324	226	250	−501	110	377	038	065	089	342	102	272	−627	502	411	600					
TMNO	404	466	431	511	345	331	485	637	−262	039	292	−249	−239	376	−069	−107	099	393	049	144	−525	686	351	300	350				
TTIO	102	195	241	238	052	095	274	231	−166	−066	242	−065	−128	158	024	−007	014	201	−024	007	−350	499	217	−006	150	385			
TKAO	136	−014	−017	−017	052	158	−066	−175	−133	227	018	−234	666	−131	004	049	−153	036	183	285	−224	−042	192	−015	260	−186	−113		
TPHO	173	193	144	198	085	269	239	380	−118	054	109	−216	−082	073	344	376	387	734	249	357	−471	392	409	278	275	333	224	043	

NOTE: Abbreviations for variable names are as in the preceding; TSIO to TPHO are for total elemental oxides in the order of SiO$_2$, Fe$_2$O$_3$, Al$_2$O$_3$, CaO, MgO, MnO$_2$, TiO$_2$, K$_2$O and P$_2$O$_5$.

To facilitate distinction of the different degrees of correlation, the table
is transformed into a figure by dividing the range of correlation coefficients into five grades, each designated by a specific pattern as shown
in the legend of Fig. 5.28(A). In Figure 5.28(B) areas are numbered, and
in the discussion that follows these numbers are used.

(1) Mutual Correlation between Base Status Characters

The correlation coefficients in this area are generally high and
positive. Ex-(Ca + Mg) is very highly correlated with Ex-Ca, CEC,
Ex-Mg, and Available Silica. It is also correlated highly with pH and
Ex-K, and moderately with Ex-Na. CEC is highly correlated with
Ex-Ca, Ex-Mg, and Available Silica, in addition to Ex-(Ca + Mg);

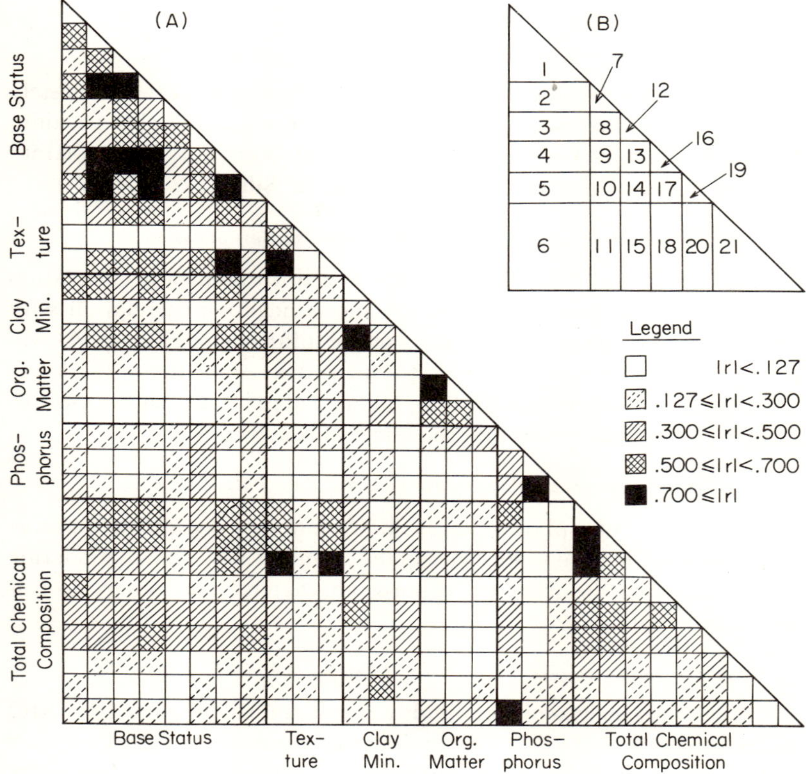

FIG. 5.28 PATTERNIZED EXPRESSION OF THE CORRELATION
MATRIX GIVEN IN TABLE 5.24

and moderately correlated with Ex-K. Correlation between Available Silica and Ex-Ca is also very high.

The lowest correlations in this area are seen between Ex-Ca and Ex-Na, pH and Ex-Mg, Available Silica and Ex-Na, and pH and Ex-Na.

(2) Correlation between Textural Composition and Base Status

Except for pH, characters related to base status are moderately to highly correlated with Sand and Clay. Silt has no significant correlation with any base status character. Sand shows negative and Clay positive correlation with the base status characters.

Although the sign is opposite, Clay and Sand show the highest correlation, with CEC, followed by that with Ex-(Ca+Mg), Ex-Mg, Ex-Ca, and Ex-K. Available Silica shows a moderate correlation and Ex-Na the lowest.

(3) Correlation between Clay Mineral Composition and Base Status

Except for Ex-Na, characters related to base status are moderately to highly correlated with 7 Å and 14 Å minerals in the clay fraction. The 10 Å minerals, however, show low correlation with all the characters, the highest being the one with Available Silica.

(4) Correlation between Organic Matter Status and Base Status

TC, TN, and NH_3-N show only low to insignificant correlation with base status characters. The pH shows negative correlation with organic matter presumably because of the influence of climate. As will be seen in the last part of this chapter, the drier the climate the lower the organic matter content and the higher the pH. CEC shows positive correlations probably through the effect of texture.

(5) Correlation between Phosphorus Status and Base Status

Generally, phosphorus status shows low to insignificant positive correlation with base status characters. Only Ex-K shows moderate correlation with the three phosphorus characters. This observation is difficult to explain.

(6) Correlation between Total Chemical Composition and Base Status

Base status characters are highly correlated negatively with TSIO and positively with TFEO. This implies that the stronger the weathering and leaching, the more silica and the less iron in the soil. The latter relation between weathering and iron content contradicts our experience that in cases of extreme weathering iron is residually accumulated in the form of pisoliths and lateritic nodules. But the correlation we have seen here deals only with paddy soils, the main occurrence of which is in recent alluvial sediments.

High correlations are also found between TALO and CEC, TCAO and pH, TMNO and Ex-(Ca+Mg), and TMNO and Available Silica. But TKAO and TTIO show insignificant to low correlations with all the base status characters. TPHO is similar to these two, except in the case of Available Silica.

Although TMGO and TMNO are moderately correlated with base status characters, TCAO shows relatively low correlation. This may be owing to the occurrence of soils containing a high amount of calcium in the sand fraction.

It is interesting to note that total silica and available silica are highly negatively correlated and that total potash and exchangeable potash are not significantly correlated.

(7) Mutual Correlation among Textural Separates

A very high negative correlation is found between Sand and Clay. Silt is moderately negatively correlated with Sand, but its correlation with Clay is insignificant.

(8) Correlation between Clay Mineral Composition and Textural Composition

Correlation coefficients in this area are all low. The 7 Å minerals are correlated positively with Sand and negatively with Clay. The 10 Å minerals show a positive correlation with Clay, which is the highest in this area, and a negative correlation with Sand. Silt shows a positive correlation only with 10 Å minerals.

(9) Correlation between Organic Matter Status and Textural Composition

Again all the correlations are low. The fact that Sand is negatively and Clay is positively correlated with all the three characters, however, indicates a certain control exerted by texture on the accumulation of organic matter, especially of TC. This point will be further elaborated later.

(10) Correlation between Phosphorus Status and Textural Composition

All the correlations are low to insignificant; in particular, those of available phosphorus (Bray-P and HCl-P) with texture are insignificant.

(11) Correlation between Total Chemical Composition and Textural Composition

Sand and Clay are highly correlated with TSIO, TFEO, and TALO. TSIO shows a positive correlation with Sand and a negative correlation with Clay. The reverse is true for TFEO and TALO. The rest of the correlations are low to insignificant, except for a moderate negative correlation between TMGO and Sand.

(12) Mutual Correlation between Clay Mineral Species

The correlations are all negative. A very high negative correlation between 7 Å and 14 Å minerals is noteworthy.

(13) Correlation between Organic Matter Status and Clay Mineral Composition

In this area only the 10 Å minerals show low negative correlations with TC and NH_3-N.

(14) Correlation between Phosphorus Status and Clay Mineral Composition

Low negative correlation is seen between 7 Å minerals and total and available phosphorus, and low positive correlation between 10 Å minerals and available phosphorus.

(15) Correlation between Total Chemical Composition and Clay Mineral
*   Composition*

Many of the correlations are low to insignificant. A high positive correlation between TKAO and 10 Å minerals is noteworthy. The 10 Å minerals contain about 10 percent potash in their interlayer spaces. Another high, but negative, correlation is seen between 7 Å minerals and TMGO. Unlike many of the 14 Å minerals, 7 Å minerals are devoid of Mg in their octahedral layers, thus high 7 Å minerals tend to be accompanied by a low TMGO content.

(16) Mutual Correlation between the Characters Related to Organic Matter Status

A very high positive correlation is seen between TC and TN, in fact, the highest of all the correlations. As stated previously, the C/N ratio of the soil converges to a narrow range (10–12), which suggests the high correlation. NH_3-N is also highly correlated with TC and TN.

(17) Correlation between Phosphorus Status and Organic Matter Status

Total phosphorus content is moderately correlated with TC, TN, and NH_3-N, while available phosphorus content is not.

(18) Correlation between Total Chemical Composition and Organic Matter Status

TPHO is moderately correlated with organic matter. TALO is positively and TSIO is negatively correlated with organic matter, which may be explained in terms of texture. Others are insignificant, except for a low negative correlation between TKAO and NH_3-N.

(19) Mutual Correlation between Phosphorus Status Characters

Available phosphorus contents determined by two different methods (Bray and HCl) are very highly correlated, while correlation between TP and available phosphorus is moderate. It appears that the more drastic the method of available phosphorus determination, the higher the correlation.

(20) Correlation between Total Chemical Composition and Phosphorus Status

Naturally, total phosphorus contents determined with chemical method and X-ray fluorescence method are highly correlated. But the correlation is not as high as expected because of the discrepancy of the two data, as was mentioned in the preceding section.

Except for TKAO, all the elemental oxides are more or less correlated with TP. The highest, negative, correlation is with TSIO. The other elements show low to moderate positive correlation.

Available phosphorus is mostly not correlated with elemental oxides, but here again the one determined by a more drastic method (HCl-P) tends to be relatively more highly correlated.

(21) Mutual Correlation between Total Contents of Elemental Oxides

Very high negative correlations exist between TSIO and TFEO and between TSIO and TALO. TSIO shows high negative correlation with TMGO and TMNO, too, and moderate negative correlation with TPHO, TCAO and TTIO. A low negative correlation is seen between TSIO and TKAO.

TFEO is similar to TSIO but its correlations are all positive. TALO has a high correlation with TFEO, but shows generally lower correlations with other elements. TCAO has a high correlation with TMGO, but its correlations with other elements are not high. TMGO and TMNO behave similarly, showing high correlations with TSIO and TFEO. The highest correlation of TTIO is with TFEO. TKAO has low to insignificant correlations with other elements. Naturally, TPHO behaves similarly to TP.

From the preceding discussion on the correlation matrix, some important general remarks may be made:

(a) Base status characters are highly correlated not only among themselves but also with textural composition, clay mineralogy, and part of the total chemical composition. Silt, 10 Å minerals, TTIO and TKAO are exceptional, showing only low to insignificant correlations with base status characters.

(b) Of the mutually correlated characters referred to in (a), Sand content in the soil, 7 Å mineral content in the clay fraction, and TSIO content in the soil are negatively correlated with most of the other characters, though their mutual correlations are positive.

(c) Characters representing organic matter are not highly correlated with any of the character groups, though their mutual correlations are high.

(d) The same can be said of the characters related to available phosphorus status.

Regression Analysis

As just stated, CEC and Clay are highly correlated, $r = 0.791$. Regression of CEC upon Clay can be formulated as follows:

$$CEC = 0.44 \ Clay + 1.72 \tag{1}$$

This equation (1) indicates that every 1 percent of clay contributes 0.44 me of CEC, or 100 g clay has, on average, 44 me of CEC. The coefficient of determination for this regression is $r^2 = 0.63$.

To further elaborate the CEC-Clay relationship, multiple regression of CEC on different clay species was studied. The content of each clay species in the soil was computed from total clay content and relative proportions of different clay species. Thus 7 Å mineral content (CL 7), 10 Å mineral content (CL 10), and 14 Å mineral content (CL 14) of each soil were obtained. Correlation coefficients between CEC, CL 7, CL 10, and CL 14 for 410 sample soils were as follows:

	CEC	CL 7	CL 10
CL 7	.150		
CL 10	.067	.164	
CL 14	.915	−.011	−.061

The correlation matrix indicates that CEC is correlated very highly with CL 14, only slightly with CL 7, and insignificantly with CL 10. The mutual correlation between the three clay contents are low to insignificant. Therefore, multicolinearity is not a problem in this multiple regression analysis.

The following 3 regression equations were obtained in steps:

$$CEC = 7.15 + 0.68^{**} CL\ 14 \qquad\qquad (R = 0.915) \tag{2}$$
$$CEC = 4.66 + 0.68^{**} CL\ 14 + 0.15^{**} CL\ 7 \qquad (R = 0.929) \tag{3}$$
$$CEC = 3.83 + 0.69^{**} CL\ 14 + 0.13^{**} CL\ 7$$
$$+ 0.20^{**} CL\ 10 \qquad\qquad (R = 0.934) \tag{4}$$

$$^{**}\text{Significant at } \alpha = 0.01$$

The partial regression coefficients for all the three clay species are highly significant, though the increase in the coefficient of determination is not great in the second and third steps.

The contribution of organic matter (in terms of TC) to CEC was checked by using TC as the fourth independent variable. The following equation was obtained:

$$CEC = 3.58 + 0.68**CL\ 14 + 0.12**CL\ 7$$
$$+ 0.20**CL\ 10 + 0.34\ TC \qquad (R = 0.935)\ (5)$$

The partial regression coefficient for TC in this equation is not significant even at $\alpha = 0.05$, indicating that organic matter makes no statistically significant contribution to CEC. We do not know at this moment whether this result is generally applicable to tropical soils or if it is only applicable to tropical paddy soils.

Each of the partial regression coefficients in equation (4) has the following confidence limit at $\alpha = 0.05$:

$$0.66 \leqslant b_1 \leqslant 0.71 \qquad \text{for CL 14}$$
$$0.10 \leqslant b_2 \leqslant 0.17 \qquad \text{for CL\ \ 7}$$
$$0.13 \leqslant b_3 \leqslant 0.27 \qquad \text{for CL 10}$$

The ranges appear resonable in comparison with the generally accepted values of CEC for various clay mineral species. The 14 Å minerals consist of montmorillonite, vermiculite, and Al-interlayered vermiculite-chlorite intergrades, of which montmorillonite is dominant. The 10 Å minerals are mostly illite and clay size micas. The 7 Å minerals include kaolinite and meta-halloysite.

Other attempts at multiple regression analyses taking Available Silica, Ex-K, and NH_3-N as dependent variables were not successful, because they had low coefficients of determination.

Effect of Climate, Relief, and Soil Texture on Organic Matter Status

It has been noted that the content of organic matter in the soil is controlled to some extent by such factors as climate, local relief or swampiness of the terrain, and soil texture. To ascertain this aspect of correlation in a quantitative manner we used Hayashi's theory of quantification No. 1.

We earlier established climatic regions (see chapter 2) for the entire region of tropical Asia. Referring to the map of climatic regions, a climatic class code is given to each sample. The samples from the Philippines were not included in this study because of the complexity of the climatic regions of the archipelago. Altogether 356 samples from six climatic regions, I through V and VII, were used in this study. They bear a climatic class code, which is the region number.

Local relief has two codes, 1 and 2; 1 is for ordinary nonswampy terrain, and 2 is for permanently swampy terrain. The allocation of codes was based on our field observations.

As a preparatory step, mean TC, TN and NH_3-N contents were computed for each of the combinations between climatic class and local relief class (see Table 5.25). The numbers of samples for climatic classes 3 and 4 are small, and no samples occur in swampy terrain in these two climatic classes. Fewer samples are located in swampy conditions in the alternately wet and dry climate (classes 4, 5, and 7) than in the permanently humid climate (classes 1, 2, and 3).

From Table 5.25, it is clear that the amount of organic matter accumulated is higher in the wetter climate and that relief affects the amount of organic matter conspicuously.

The effect of soil texture is shown in Table 5.26. In the table, textural classes are coded as follows: 1-HC, 2-LiC, 3-SiC, 4-SC, 5-CL, 6-SiCL, 7-SCL, 8-L and SiL, 9-SL, and 10-LS and S. The TC, TN, and NH_3-N contents are generally higher as soil texture becomes finer. In an Indonesian ando soil with a clay loam texture (code 5), an extraordinarily high amount of organic matter is found—that is, 5.6 percent TC; 0.59 percent TN; and 36.2 mg/100 g of NH_3-N. The mean values

TABLE 5.25   ORGANIC MATTER STATUS AS BROKEN DOWN BY
CLIMATE AND RELIEF

CLIMATE	RELIEF	No. OF SAMPLES	TC (IN PERCENT)	TN (IN PERCENT)	NH_3-N (mg/100g)
	1	39	1.29	0.10	11.7
1	2	11	5.50	0.47	26.0
	Mean	50	2.22	0.18	14.9
	1	28	1.42	0.13	11.3
2	2	14	3.45	0.28	15.4
	Mean	42	2.10	0.18	12.6
	1	7	2.21	0.22	20.6
3	2	0	—	—	—
	Mean	7	2.21	0.22	20.6
	1	10	1.23	0.12	3.7
4	2	0	—	—	—
	Mean	10	1.23	0.12	3.7
	1	151	1.05	0.09	4.3
5	2	6	2.97	0.24	6.4
	Mean	157	1.12	0.10	4.4
	1	83	0.86	0.09	3.8
7	2	7	2.34	0.22	11.8
	Mean	90	0.98	0.10	4.4

TABLE 5.26   ORGANIC MATTER STATUS AS BROKEN DOWN BY TEXTURE

TEXTURE	NO. OF SAMPLES	TC (IN PERCENT)	TN (IN PERCENT)	NH_3-N (mg/100g)
1	137	1.93	0.16	8.90
2	63	1.24	0.11	8.23
3	20	1.40	0.14	6.77
4	14	1.15	0.10	4.94
5	35	1.32	0.13	7.01
		(1.20)	(0.11)	(6.15)
6	13	0.99	0.11	5.62
7	18	0.89	0.08	4.46
8	13	0.70	0.07	3.62
9	28	0.44	0.05	3.07
10	15	0.58	0.06	4.60

NOTE: Figures in parentheses represent the means calculated after excluding an Indonesian ando soil.

of the class are considerably lower when this soil is excluded, as shown in parentheses.

All three factors, that is, climate, relief, and texture, were taken into consideration in explaining the organic matter status. The overall mean values of the criterion variables for 356 samples (excluding Philippine soils) are as follows:

	mean	standard deviation
TC (percent)	1.38	1.34
TN (percent)	0.124	0.112
NH_3-N (mg/100g)	6.15	7.03

The seven climatic divisions were reduced to four by combining those of geographical proximity (see Table 5.27). This way the number of samples per category was increased. Texture was also recoded into four: 1 for HC; 2 for LiC, SiC, and SC; 3 for CL, SiCL, and SCL; and 4 for L, SiL, SL, LS, and S. The model for TC is as follows:

$$TC \text{ (percent)} = \text{climate (four categories)} + \text{relief (two categories)} + \text{texture (four categories)}$$

Numeric values assigned to each category of the three items (or variables) are as shown in Table 5.27. Multiple correlation coefficients (R) and partial correlation coefficients for each variable in the model are given in Table 5.28.

The R values are not sufficiently high to justify the use of the

TABLE 5.27   NUMERICAL VALUES ASSIGNED TO EACH CATEGORY OF
THREE QUALITATIVE VARIABLES TO ESTIMATE
ORGANIC MATTER STATUS

VARIABLE	CATEGORY	TC	TN	NH$_3$-N
Climate	1 + 3	0.453	0.033	7.481
	2	0.316	0.025	4.164
	4 + 5	−0.044	−0.008	−2.151
	7	−0.352	−0.018	−2.690
Relief	1	−0.251	−0.020	−0.746
	2	2.099	0.171	6.240
Texture	1	0.328	0.024	0.453
	2	−0.089	−0.006	0.401
	3	0.081	0.012	0.613
	4	−0.745	−0.062	−2.524

TABLE 5.28   PARTIAL AND MULTIPLE CORRELATION COEFFICIENTS
BETWEEN ORGANIC MATTER STATUS AND CLIMATE, RELIEF,
AND TEXTURE

PARTIAL CORR.	TC	TN	NH$_3$-N
Climate	0.274	0.215	0.611
Relief	0.596	0.568	0.395
Texture	0.356	0.333	0.219
MULTIPLE CORR.	0.706	0.674	0.725

equations for prediction. The partial correlation coefficients indicate that the relief factor is most relevant to the TC and TN contents, while climate contributes most to the amount of mineralizable nitrogen. The effect of climate on TC and TN seems to be relatively minor. Soil texture is of moderate importance for determining the TC and TN levels, but of minor importance for NH$_3$-N.

As it was anticipated from field observations that texture is more important in determining the organic matter level in the drier climate, analysis was carried out after dividing the samples into two—one group containing the samples which occur in the permanently humid climate (codes 1, 2, and 3), and the other containing those occurring in the wet and dry climate (codes 4, 5, and 7). The results are shown in Table 5.29, in terms of partial and multiple correlation coefficients.

TABLE 5.29   PARTIAL AND MULTIPLE CORRELATION COEFFICIENTS
BETWEEN ORGANIC MATTER STATUS AND RELIEF AND TEXTURE
AFTER STRATIFYING THE SAMPLES BY CLIMATE

PERMANENTLY HUMID CLIMATE (NUMBER OF SAMPLES = 99)

Partial Corr.	TC	TN	NH_3-N
Relief	0.648	0.625	0.410
Texture	0.393	0.354	0.367
Multiple Corr.	0.690	0.661	0.509

ALTERNATELY WET AND DRY CLIMATE (NUMBER OF SAMPLES = 257)

Partial Corr.	TC	TN	NH_3-N
Relief	0.499	0.478	0.321
Texture	0.522	0.479	0.238
Multiple Corr.	0.674	0.637	0.407

The multiple correlation coefficients are low for NH_3-N, regardless of climatic zone. Relief is even more relevant to TC and TN in the permanently humid climate, while texture plays as important a role as relief for TC and TN in the wet and dry climate.

6

Soil Material Classification of Paddy Soils in Tropical Asia

As stated in chapter 5, the material characteristics of tropical Asian paddy soils vary from one country to another, and even from one region to another within each country. This variability occurs because most paddy soils are alluvial soils. The two following statements are relevant to the consideration of the nature of alluvial soils, and, accordingly, of paddy soils (Kyuma *et al.*, 1974).

First, alluvial soils are by definition not markedly different from the most recent geological sediments laid down in the lowest part of a terrain by fluvial or marine action. As they have undergone little or no pedogenetical change, no specific morphology has developed on their profiles. Thus, the nature of alluvial soils is directly governed by the nature of the parent sediments.

Second, variability of soil is by far the greatest in alluvial soils, of all the soil groups. It is conditioned by the geology and the degree of weathering in the catchment area and/or the milieu of sedimentation. At one extreme, fresh volcanic ejecta are laid down in alluvia (for example, on the island of Java), while at the other extreme there are such cases as those mentioned in the Seventh Approximation (Soil Survey Staff, 1960) where: "Chemically and mineralogically, the materials in an oxic horizon may be indistinguishable from materials at comparable depths in Entisols." It should be further noted that the range of variability is wider in the tropics than in the temperate and cold regions.

In view of this, soil material is of paramount importance in the classification of soils in alluvial lowlands in the tropics. This is especially true for such cultivated soils as paddy soil, because soil material is directly connected with soil fertility.

In the present U.S. soil classification system (Soil Survey Staff, 1960, 1967), material, in terms of mineralogy and texture, is used as a differentiating characteristic for the "family" category, and this differentia is carried over to the lowest category, "soil series". In actual soil classification, however, the family category is the least elaborated, and soil material is often not taken into proper consideration. In the case of residual soils in uplands, the geology or petrology conveniently may be used instead of the soil material. But this is not possible for alluvial soils. Therefore, the significance of soil material factor in alluvial soil studies is great.

In this chapter we deal with this problem of soil material as the prerequisite for a more rational classification of alluvial soils.

6.1 DATA AND METHODS USED

The data for the plow layer soil of the same 410 samples described and analyzed in chapter 5 were used.

In the classification of soil materials both chemical and mechanical characteristics should be taken into consideration. At the same time, for a method to be practically useful the data should not be too many. Therefore, we decided to take two sets of data, total chemical composition and mechanical composition.

The method of classification may be described under two headings; (a) setting up of soil material classes, and (b) placing samples in appropriate classes.

(a) Setting Up of Soil Material Classes

The numerical taxonomic method proposed by Sokal and Sneath (1963) is useful for this purpose. Numerical taxonomy may be defined as "the numerical evaluation of the affinity or similarity between taxonomic units and the ordering of these units into taxa on the basis of their affinities." It was originally proposed for general or natural classification of such objects as plants, insects, microbes, etc. In this study, however, the method is used to create groups on the basis of the similarity of a limited number of characters which are relevant to the material features of the soil.

The actual procedure of numerical taxonomy consists of the following steps: standardization of the data to make them dimensionless, computation of between-sample similarity coefficients, sorting or clustering, formulation of a dendrogram.

Two sets of data, total chemical composition and mechanical composition, are used in this study. Since the twelve characters, that is, the content of nine elemental oxides and three textural separates, are correlated with each other at varying degrees (Table 5.24 of chapter 5), the relative weight given to chemical and mechanical characteristics is not necessarily proportionate to the number of data used. In this situation it would be better to use fewer mutually independent compound characters that can be extracted from the original characters by means of principal component analysis (PCA). (This method will be explained briefly in chapter 7.)

Starting from the correlation matrix of the 12 characters for the 410 sample soils, principal components were extracted, the eigenvalues and eigenvectors of which are given in Table 6.1. The eigenvalues of the first three components are greater than 1, and the cumulative percentage of eigenvalues for the three components is about 70 percent of total variance. In Table 6.2 correlation coefficients (or factor loadings) between principal components and characters are given, together with the communalities for the first three components. From the communality figures, we can see that the greater part of the information from the original data is represented by the three principal components, except in the cases of Silt, TiO_2, K_2O, and P_2O_5.

The first principal component represents the fundamental character of the material as determined by the three major elemental oxides, SiO_2, Fe_2O_3, and Al_2O_3, and texture. Alkaline earth bases, MnO_2, TiO_2 and P_2O_5 contribute moderately to this component. A high positive score is because of clayey texture and/or low silica, and high iron and alumina contents; a high negative score would indicate a sandy and/or siliceous nature.*

The second principal component is related to texture and mineral reserve. A high positive score is expected when a soil contains a large amount of volcanic sands of basic composition, and a high negative score is expected when a soil is derived from silty to clayey sediments

---

*Formulae for computation of principal component scores (PCS) for use in soil material classification have been derived as follows:

$$PCS(1) = 1/\sqrt{4.98} \times (-0.34X_1 + 0.13X_2 + 0.33X_3 - 0.43X_4 + 0.39X_5 + 0.39X_6 + 0.17X_7 + 0.29X_8 + 0.28X_9 + 0.19X_{10} + 0.06X_{11} + 0.22X_{12})$$

$$PCS(2) = 1/\sqrt{1.74} \times (0.42X_1 - 0.42X_2 - 0.24X_3 + 0.00X_4 + 0.19X_5 - 0.18X_6 + 0.35X_7 + 0.11X_8 + 0.36X_9 + 0.27X_{10} - 0.35X_{11} + 0.26X_{12})$$

$$PCS(3) = 1/\sqrt{1.51} \times (0.15X_1 + 0.24X_2 - 0.33X_3 - 0.07X_4 - 0.14X_5 - 0.05X_6 + 0.50X_7 + 0.45X_8 - 0.12X_9 - 0.35X_{10} + 0.43X_{11} + 0.15X_{12})$$

where

$$X_1 = (x_1 - 33.88)/25.99 \qquad X_2 = (x_2 - 27.67)/13.66$$
$$X_3 = (x_3 - 38.44)/21.65 \qquad X_4 = (x_4 - 72.13)/11.49$$
$$X_5 = (x_5 - 5.96)/\,3.73 \qquad X_6 = (x_6 - 16.38)/\,6.97$$
$$X_7 = (x_7 - 1.42)/\,1.96 \qquad X_8 = (x_8 - 0.92)/\,0.77$$
$$X_9 = (x_9 - 0.13)/\,0.12 \qquad X_{10} = (x_{10} - 1.15)/\,0.60$$
$$X_{11} = (x_{11} - 1.83)/\,1.27 \qquad X_{12} = (x_{12} - 0.13)/\,0.08$$

$x_1 - x_{12}$ are the original variables used for extracting the principal components (see Table 6.1, for the variable numbers).

TABLE 6.1 EIGENVALUES AND EIGENVECTORS FOR THE FIRST FOUR PRINCIPAL COMPONENTS

PRINCIPAL COMPONENT	1	2	3	4
Eigenvalue	4.981	1.737	1.508	0.908
Cumulative percent of Total Variance	41.5	56.0	68.6	76.1
Eigenvector				
1 Sand	−0.338	0.420	0.153	0.230
2 Silt	0.132	−0.420	0.239	−0.413
3 Clay	0.325	−0.241	−0.334	−0.017
4 SiO_2	−0.433	0.002	−0.069	−0.124
5 Fe_2O_3	0.389	0.186	−0.135	−0.012
6 Al_2O_3	0.387	−0.179	−0.051	0.173
7 CaO	0.169	0.349	0.495	−0.301
8 MgO	0.289	0.108	0.447	−0.111
9 MnO_2	0.278	0.355	−0.123	−0.230
10 TiO_2	0.188	0.270	−0.347	0.247
11 K_2O	0.061	−0.352	0.425	0.597
12 P_2O_5	0.217	0.256	0.147	0.397

TABLE 6.2 FACTOR LOADING MATRIX AND COMMUNALITY FOR THE FIRST THREE PRINCIPAL COMPONENTS

PRINCIPAL COMPONENT	1	2	3	COMMUNALITY
Sand	−0.754	0.554	0.188	0.910
Silt	0.294	−0.553	0.294	0.479
Clay	0.726	−0.318	−0.410	0.796
SiO_2	−0.965	0.002	−0.085	0.939
Fe_2O_3	0.868	0.245	−0.166	0.840
Al_2O_3	0.864	−0.236	−0.063	0.805
CaO	0.376	0.460	0.608	0.723
MgO	0.644	0.142	0.549	0.737
MnO_2	0.620	0.468	−0.151	0.625
TiO_2	0.419	0.356	−0.426	0.484
K_2O	0.137	−0.465	0.522	0.507
P_2O_5	0.484	0.338	0.180	0.381

containing a high amount of illite and fine micas. But when soil texture is extremely sandy or clayey, and calcium, manganese, and potash contents are moderate, the scores are almost solely determined by the texture.

The third principal component is mainly related to the base status of the soil. Clay and TiO_2 contents contribute moderately to this component. A high positive score results from high alkaline earth base and potash contents, especially when clay and TiO_2 contents are low; conversely a high negative score is expected for a soil with high contents of unsaturated (base-poor) clays and/or TiO_2.

As these three principal components are mutually independent and represent different aspects of such important material characteristics as siliceousness, texture, weatherable mineral reserve, and base status, their scores can be taken as the criteria of soil material classification, each one being given equal weight. Taxonomic distances between all pairs of the samples were computed by the formula (Sokal and Sneath, 1963):

$$d_{jk} = \left[\frac{\sum_{i=1}^{p} (x_{ij} - x_{ik})^2}{p} \right]^{\frac{1}{2}}$$

and adopted as the similarity coefficients for numerical taxonomy. In this formula x_{ij} is the standardized ith principal component score of the jth sample, and p is the number of variables used, here $p = 3$.

The weighted pair-group method is used for clustering, allowing the two mutually nearest operational taxonomic units to join in one clustering cycle. The final result of the clustering process is illustrated by a dendrogram (Fig. 6.1).

(b) Placing Samples in Appropriate Classes

Using the dendrogram we can set up soil material classes. Once the classes are established, the next step in classification is to objectively place a new sample with the appropriate data in one of these classes. Discriminant functions appear to serve this purpose effectively.

To illustrate this method let us consider a case of two variables and several groups (for details see Okuno, T. *et al.*, 1971). Suppose a sample point is specified in two-dimensional space, on an X_1–X_2 plane. The general procedure of discrimination is first to compare the distances from the sample to the gravity centers of the classes (or groups), and then to put the sample into the nearest class. Mahalanobis' generalized distance, which is a squared distance in the principal component coordinates, is used. Instead of computing the distances for each sample, the locus of equidistance points from the two groups to be compared is found. If the two groups have an equal variance-covariance matrix, the locus becomes a straight line, and a sample point on one side of the line belongs to the class whose center is on the same side. To simplify the procedure further, the sample point is projected on a straight line perpendicularly crossing the locus of equidistance points. The straight

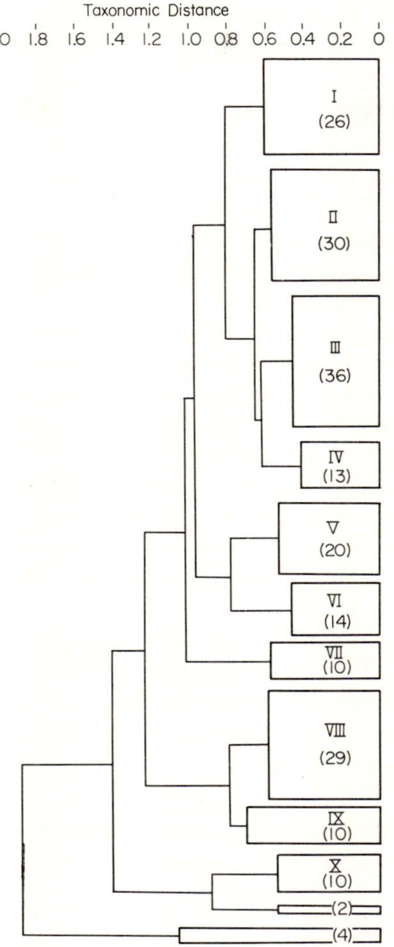

FIG. 6.1   SCHEMATIC DENDROGRAM SHOWING THE RELATION OF THE TEN SOIL MATERIAL CLASSES

line is generally expressed as $z = a_0 + a_1 x_1 + a_2 x_2$, where a_0 is determined to make $z = 0$ when x_1 and x_2 represent the points on the locus. The value of z assumes, therefore, a positive or a negative sign depending on which side of the locus the sample point is located. This straight line z is the discriminant function. If such discriminant functions are derived in advance for all pairs of classes, classification of a new sample is simply a matter of checking the sign of z for each pair of classes.

6.2 RESULTS OF CLASSIFICATION

Because of the limits of the computer program of numerical taxonomy 50 percent of the samples were randomly selected by the "Sample" program in SPSS, Statistical Package for Social Sciences (SPSS-Kyoto Version, Data Processing Center of Kyoto University).

Taxonomic distances were computed for all pairs of the 204 samples, using the three principal component scores derived from the twelve data of mechanical and total chemical compositions. By sorting with the weighted pair-group method, a dendrogram, shown schematically in Figure 6.1, was prepared.

On the dendrogram a straight line intersecting the distance axis at 0.60 was drawn, and the clusters formulated before the threshold values were numbered as I through X in roman numerals from the top to the bottom of the dendrogram. The value of 0.60 was chosen arbitrarily to produce an appropriate number of clusters. For class IX the threshold distance value was shifted to 0.79 to save a group of 10 samples and in consideration of the relatively large distance of the group from adjacent classes. Because 6 samples at the bottom of the dendrogram were omitted from classification, 198 samples were classified.

The means and standard deviations of the three principal component scores (PCS) were calculated for each of the ten classes, as shown in Table 6.3. Standard deviations are generally small, indicating fair homogeneity. Only in a few cases does it exceed 0.5, or a half the standard deviation for the whole sample.

TABLE 6.3   MEANS AND STANDARD DEVIATIONS OF THE THREE PRINCIPAL COMPONENT SCORES FOR THE TEN SOIL MATERIAL CLASSES

CLASS	PCS (1) Mean	PCS (1) S.D.	PCS (2) Mean	PCS (2) S.D.	PCS (3) Mean	PCS (3) S.D.
I	0.807	0.356	0.126	0.703	−1.029	0.353
II	−0.674	0.354	−0.614	0.365	−0.177	0.514
III	0.402	0.325	−1.204	0.340	−0.021	0.497
IV	0.449	0.376	−0.196	0.243	0.196	0.271
V	−0.385	0.668	0.495	0.401	0.717	0.378
VI	1.283	0.328	0.971	0.402	0.006	0.372
VII	0.648	0.637	−1.025	0.238	1.380	0.270
VIII	−1.946	0.351	0.554	0.465	−0.326	0.254
IX	−0.505	0.335	1.370	0.535	−0.581	0.560
X	0.473	0.416	1.782	0.462	1.693	0.443

To see the basis of differentiation of the classes, the pattern of the three mean scores are shown in Figure 6.2. The characteristics of each group can be described with respect to each of the PCS's, but this will be done later, when we have all 410 samples classified.

Having established ten soil material classes, we can now derive the discriminant functions. To derive linear discriminant functions, we have to assume equality of variance-covariance matrices for the ten classes, though usually this assumption is not strictly held.

The coefficients for the discriminant function of each of $\binom{10}{2} = 45$ pairs and the discriminant efficiency (that is, the Mahalanobis' distance D^2 between the two groups concerned) are given in Table 6.4. The probability of error, that is, misclassification, can be computed by $E = D/2$, which is compared with a certain probability point of the normal distribution; for example, where $E = 2.00$, the probability of error is expected to be 2.28 percent.

The discriminant efficiency is lowest between classes III and IV, and next lowest between IV and V, II and III, and II and IV, the error probability for these pairs being 12.3, 7.6, 6.9 and 6.6 percent, respectively. When the same 198 samples that make up the ten classes were reclassified by means of discriminant functions, 22 samples or about 11 percent were misclassified. Discriminant scores, however, indicated that most of the misclassification cases occurred right at the margin of the classes concerned. Since this kind and degree of error is inevitable, we decided to use the discriminant functions for further steps.

All the soil samples including those unclassified in the dendrogram, were classified into one of the ten classes by means of discriminant functions. The result is summarized in Table 6.5, which shows the distribution of the samples from each country among the different classes.

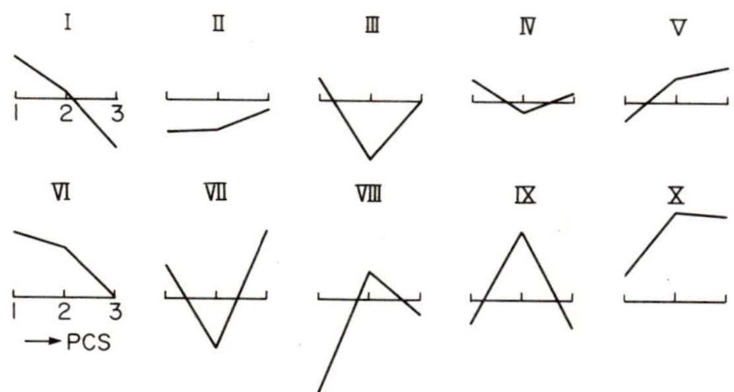

FIG. 6.2   PATTERNS OF THE MEAN PRINCIPAL COMPONENT SCORES (PCS) FOR THE TEN SOIL MATERIAL CLASSES (I ~ X)

TABLE 6.4   COEFFICIENTS AND DISCRIMINANT EFFICIENCY (D.E.) OF THE DISCRIMINANT FUNCTION FOR EACH PAIR OF THE TEN SOIL MATERIAL CLASSES

	I ∼ II	I ∼ III	I ∼ IV	I ∼ V	I ∼ VI	I ∼ VII	I ∼ VIII	I ∼ IX	I ∼ X
a_0	−2.698	−0.949	−4.281	−2.513	2.158	4.421	7.207	1.404	12.028
a_1	8.764	2.390	2.118	7.059	−2.813	0.940	16.294	7.768	1.984
a_2	3.738	6.715	1.628	−1.862	−4.265	5.816	−2.165	−6.285	−8.358
a_3	−5.025	−5.947	−7.225	−10.298	−6.101	−14.205	−4.151	−2.643	−16.046
D.E.	20.025	15.888	10.133	27.080	11.257	41.054	48.692	19.190	58.164

	II ∼ III	II ∼ IV	II ∼ V	II ∼ VI	II ∼ VII	II ∼ VIII	II ∼ IX	II ∼ X	III ∼ IV
a_0	1.749	−1.583	0.185	4.856	7.119	9.905	4.102	14.726	−3.332
a_1	−6.374	−6.646	−1.705	−11.576	−7.824	7.530	−0.996	−6.780	−0.272
a_2	2.978	−2.109	−5.599	−8.002	2.078	−5.903	−10.022	−12.095	−5.087
a_3	−0.921	−2.200	−5.272	−1.076	−9.180	0.875	2.383	−11.020	−1.279
D.E.	8.760	9.169	11.418	35.538	25.488	16.600	21.018	57.364	5.415

	III ∼ V	III ∼ VI	III ∼ VII	III ∼ VIII	III ∼ IX	III ∼ X	IV ∼ V	IV ∼ VI	IV ∼ VII
a_0	−1.565	3.107	5.369	8.155	2.352	12.976	1.767	6.439	8.701
a_1	4.669	−5.203	−1.450	13.904	5.378	−0.406	4.942	−4.930	−1.178
a_2	−8.577	−10.980	−0.900	−8.881	−13.000	−15.073	−3.490	−5.893	4.188
a_3	−4.351	−0.154	−8.259	1.796	3.304	−10.099	−3.072	1.124	−6.980
D.E.	21.456	28.462	12.083	48.801	40.185	62.328	8.134	11.202	11.965

	IV ~ VIII	IV ~ IX	IV ~ X	V ~ VI	V ~ VII	V ~ VIII	V ~ IX	V ~ X	VI ~ VII
a_0	11.487	5.684	16.308	4.672	6.934	9.720	3.917	14.541	2.263
a_1	14.176	5.650	−0.133	−9.872	−6.119	9.235	0.709	−5.075	3.753
a_2	−3.793	−7.913	−9.986	−2.403	7.678	−0.303	−4.423	−6.496	10.081
a_3	3.075	4.583	−8.821	4.197	−3.908	6.147	7.655	−5.748	−8.104
D.E.	38.397	21.342	32.951	20.594	20.577	20.838	13.890	18.322	33.634

	VI ~ VIII	VI ~ IX	VI ~ X	VII ~ VIII	VII ~ IX	VII ~ X	VIII ~ IX	VIII ~ X	IX ~ X
a_0	5.049	−0.755	9.869	2.786	−3.017	7.607	−5.803	4.821	10.624
a_1	19.106	10.581	4.797	15.354	6.828	1.044	−8.526	−14.310	−5.784
a_2	2.100	−2.020	−4.093	−7.981	−12.100	−14.173	−4.119	−6.192	−2.073
a_3	1.950	3.458	−9.945	10.055	11.563	−1.841	1.508	−11.895	−13.403
D.E.	63.206	21.748	23.971	69.565	59.515	40.536	16.032	66.222	36.977

NOTE: The general formula for discriminant functions is

$$Z(A \sim B) = a_0 + a_1 \cdot PCS(1) + a_2 \cdot PCS(2) + a_3 \cdot PCS(3)$$

where PCS (1) ~ PCS (3) are the values computed as in the footnote of page 126. The model is designed so as that positive Z indicates a sample belonging to class A and negative Z one belonging to class B.

SOIL MATERIAL CLASS	I	II	III	IV	V	VI	VII	VIII	IX	X	SUM
Bangladesh	1 (1.9)	14 (26.4)	12 (22.6)	0 (0)	6 (11.3)	0 (0)	18 (34.0)	2 (3.8)	0 (0)	0 (0)	53
Burma	2 (12.5)	1 (6.2)	3 (18.8)	3 (18.8)	4 (25.0)	0 (0)	2 (12.5)	0 (0)	0 (0)	1 (6.2)	16
Cambodia	4 (25.0)	3 (18.8)	2 (12.5)	0 (0)	0 (0)	0 (0)	0 (0)	6 (37.5)	1 (6.2)	0 (0)	16
India	7 (9.6)	9 (12.3)	6 (8.2)	8 (11.0)	16 (21.9)	10 (13.7)	4 (5.5)	5 (6.8)	6 (8.2)	2 (2.7)	73
Indonesia	19 (43.1)	2 (4.5)	0 (0)	0 (0)	3 (6.8)	14 (31.8)	0 (0)	0 (0)	0 (0)	6 (13.6)	44
Malaysia	0 (0)	16 (39.0)	22 (53.7)	0 (0)	0 (0)	0 (0)	0 (0)	3 (7.3)	0 (0)	0 (0)	41
Philippines	16 (29.6)	1 (1.8)	1 (1.8)	15 (27.8)	6 (11.1)	3 (5.6)	1 (1.8)	0 (0)	3 (5.6)	8 (14.8)	54
Sri Lanka	3 (9.1)	1 (3.0)	0 (0)	2 (6.1)	6 (18.2)	1 (3.0)	0 (0)	4 (12.1)	14 (42.4)	2 (6.1)	33
Thailand	9 (11.2)	21 (26.2)	20 (25.0)	1 (1.2)	1 (1.2)	3 (3.8)	0 (0)	24 (30.0)	1 (1.2)	0 (0)	80
TOTAL	61	68	66	29	42	31	25	44	25	19	410

NOTE: Figures in parentheses represent percent.

The overall distribution of the samples among the classes is generally parallel to that of the selected samples. The greatest number of samples is concentrated in class II and III, and the smallest number occurs in class X.

From the table it can be seen that the number of classes with more than 10 percent of the samples from each country is only two for Malaysia, three for Indonesia, and four for most of the other countries. The highest concentration is 54 percent of Malaysian soils occuring in class III, with more than 40 percent in class I in the case of Indonesia and in class IX for Sri Lanka. Thus soil materials show certain regionality.

The mean mechanical and total chemical composition for the samples in each group are given in Table 6.6, together with the means of such characters related to soil materials as pH, percentage base saturation (PBS), CEC, and clay mineralogical composition. Referring to Tables 6.3, 6.5 and 6.6, we can characterize each of the ten classes as follows:

Class I. Fine-textured material with moderate base status (some of pyroclastic origin); 14–7 clay with very few 10 Å minerals. (See chapter 5 for clay mineralogical composition.)

Class II. Medium-textured material with low base status (typically low humic gley soils on terraces); highly siliceous; 7 Å-clay dominant with moderate 14 and 10 Å minerals.

Class III. Fine-textured material with low base status (many such soils are derived from deltaic sediments); slightly siliceous; MgO > CaO; 7–14 clay with moderate 10 Å minerals.

Class IV. Medium-textured material with high base status (some of pyroclastic origin); 14–7 clay with few 10 Å minerals.

Class V. Coarse-textured material with moderate base status (often occurring on river levees); moderately siliceous; high potash; 14–7 clay with moderate 10 Å minerals.

Class VI. Fine-textured material with high base status (typically calcareous alluvial soils and grumusols; many such soils are of pyroclastic origin); 14–7 clay with very few 10 Å minerals.

Class VII. Medium-textured (silty) material with moderate base status (mostly of Ganges-Brahmaputra sediment origin); very high potash; 7–10–14 clay.

Class VIII. Very coarse-textured material with very low base status (strongly weathered sandy terrace or plateau materials); very highly siliceous, 7 Å-clay dominant with few 14 Å and very few 10 Å minerals.

Class IX. Coarse-textured (low silt) material with low base status (typically local alluvial sediments in acidic rock area); moderately siliceous; 7 Å-clay dominant with moderate 14 Å and very few 10 Å minerals.

Class X. Coarse-textured material with high base status (almost

TABLE 6.6 MEAN MECHANICAL AND TOTAL CHEMICAL COMPOSITION AND SOME OTHER RELATED PROPERTIES OF THE SAMPLES IN EACH MATERIAL CLASS

CLASS (No. of Samples)	I (61)	II (68)	III (66)	IV (29)	V (42)	VI (31)	VII (25)	VIII (44)	IX (25)	X* (19)	WHOLE (410)
Sand %	12.40	34.43	8.64	21.83	58.72	20.33	18.82	73.78	68.38	54.63	33.81
Silt %	22.73	34.05	35.21	33.93	19.84	25.92	49.76	15.54	10.05	25.70	27.60
Clay %	64.87	31.52	56.15	44.20	21.43	53.76	31.43	10.68	21.56	19.68	38.59
SiO_2 %	62.85	80.45	67.40	66.55	73.81	59.00	66.65	94.62	77.55	62.70	72.12
Fe_2O_3 %	9.90	3.07	5.71	7.69	4.51	11.95	6.71	0.82	5.41	6.89	5.96
Al_2O_3 %	22.12	12.78	21.53	19.25	14.58	20.98	18.31	3.45	11.87	18.41	16.38
CaO %	1.31	0.39	0.46	1.95	2.03	3.22	1.58	0.22	0.98	6.89	1.42
MgO %	0.90	0.38	0.94	1.23	0.95	1.57	1.98	0.05	0.68	2.28	0.92
MnO_2 %	0.22	0.06	0.06	0.18	0.11	0.31	0.10	0.02	0.14	0.16	0.12
TiO_2 %	1.54	0.99	1.11	1.04	0.86	1.59	0.96	0.64	1.98	0.94	1.14
K_2O %	1.04	1.85	2.67	1.96	2.99	1.16	3.59	0.40	1.31	1.49	1.83
P_2O_5 %	0.12	0.09	0.13	0.15	0.15	0.22	0.13	0.06	0.12	0.24	0.13
pH %	6.1	5.3	5.2	6.5	6.8	7.0	6.6	5.3	6.0	6.9	6.0
PBS %	94.2	59.1	81.5	100.0	103.2	106.2	98.9	65.0	83.0	105.8	85.6
CEC me/100g	31.1	12.3	20.4	26.7	13.9	34.4	15.4	4.2	10.3	15.5	18.6
7Å %	43.1	52.4	46.7	32.6	36.8	37.3	32.4	70.1	63.0	29.3*	46.4*
10Å %	5.0	19.2	20.9	12.4	20.2	3.9	28.2	5.1	8.2	12.3*	13.9*
14Å %	51.9	28.5	32.4	55.0	43.0	58.9	39.4	24.8	28.8	58.4*	39.7*

*Because four samples in group X have no crystalline minerals, the means for the other samples were taken.

exclusively of pyroclastic origin); very high alkaline earth bases; 14–7 clay with few 10 Å minerals or amorphous clay (allophane).

As is evident from the preceding, the material classification established in this study clearly reflects fertility differences between the sample soils. For example, soils with class VI material are base saturated and contain a high amount of better-quality clay. Their mineral nutrient reserves are moderate to high, and their nutrient holding capacity is also high. On the contrary, the poorest soils are those with class VIII material, in which nothing but silica is abundant.

As will be described in chapter 7, there is a correlation between the soil fertility rating and soil material composition. Because of this correlation we can expect that, if the soil material classes are set up as a basis of "family" classification and therefore of a "soil series" separation, the interpretation of each soil series would become much easier, and its usefulness in relation to the assessment of soil capability greatly enhanced.

An objection could be made against the use of material characteristics of the surface soil for differentiating "soil families." To avoid this one can take subsoil samples for material classification instead of surface soil samples. However, it has been ascertained that there is little difference in material characteristics between the surface and subsoil samples in, for example, Thai paddy soils (Charoen, 1974).

We would like to argue in favor of using the surface soil. The majority of paddy soils are alluvial soils, or Entisols and Inceptisols in the U.S. system of soil classification. According to the Soil Survey Manual (Soil Survey Staff, 1951) "Alluvial soils do not have sequences of genetically related horizons. Materials laid down by water commonly have layers that differ greatly in such characteristics as texture, but the occurrence of those layers is governed by geological processes and is accidental insofar as genetic relationships of the soils to the present environment are concerned."

Use of subsoil characteristics for differentiating soil classes at or above the soil series level is based on the following considerations: surface soil characteristics are often modified by plowing, part or all of the surface soil characteristics may be lost by erosion, and in genetically developed soils comprehension of the subsoil characteristics can allow one to make accurate inferences about the surface soil characteristics.

In the U.S. system (Soil Survey Staff, 1967) the control section for differentiating "families" or "series" without diagnostic horizons is defined as between 25 and 100 cm. But if there is no genetic relationship between the surface and subsoil layers, as in the case of alluvial soils,

                    Paddy Soils in Tropical Asia

there is no point in having such a control section. Furthermore, in these soils there are neither specific morphological characteristics to be lost by plowing, nor is there a danger of truncation by erosion.

Even consideration of plant growth factors does not justify the control section as just defined. Ishizuka and Tanaka (1963) found that more than 90 percent of rice roots are in the surface 20 cm of the soil profile. From this one can say that the importance of subsoil material characteristics is minor in the crop performance of rice.

7

Fertility Evaluation of Paddy Soils in Tropical Asia

According to the Soil Survey Manual (Soil Survey Staff, 1951) "Soil fertility is the quality that enables the soil to provide the proper compounds in the proper amounts and in the proper balance for the growth of specified plants when other factors such as light, temperature, moisture, and the physical condition of the soil, are favorable." Thus, by definition the chemical nature of a soil determines its fertility. For paddy soils the importance of soil physical properties or tilth is relatively minor because of their specific management condition of waterlogging. Therefore, soil fertility deserves much attention in attempts to evaluate the capability of paddy soils.

There have been many attempts to evaluate soil fertility in different countries or regions, such as land capability classification and the Storie Index. In all the methods so far proposed contributing factors have been only subjectively evaluated, and the final results have usually been presented in a qualitative form.

The Soil Survey Manual also says "Soil fertility is not directly observable. Thus, soil may be grouped into fertility classes only by inference." Is it then not possible to evaluate soil fertility quantitatively, and to attain a fertility classification in a more objective and reproducible way?

In an attempt to answer this challenge we have proposed a method of fertility evaluation particularly intended for paddy soils (Kyuma, 1973b, c; Kyuma and Kawaguchi, 1973) and made a few preliminary applications of that method to tropical Asian paddy soils (Kyuma and Kawaguchi, 1971, 1972a, b). In this chapter the results of our most recent study along the same lines are presented.

7.1 DATA AND METHODS USED

The data of the plow layer samples of the same 410 tropical Asian paddy soils were used. Descriptions and the results of correlation analysis of these data were given in chapter 5.

In processing these data for fertility evaluation we use two multivariate statistical methods, principal component analysis and factor analysis. (For details, see Okuno *et al.*, 1971; Asano, 1971.) The aims and procedures of these two methods are mutually related, as described later.

Principal component analysis (PCA) is often used for attaining a "parsimonious summarization of a mass of observation" (Seal, 1964). In other words, it is used to extract the hidden essence of a thing or material that is not directly measurable, making use of the correlation between various measurable variables related to it.

Given n samples, each of which is defined by p characters, they can be expressed as n points scattered in a p-dimensional space. PCA aims at reducing the p-axes to orthogonal m-axes, where $m < p$. Mathematically this is attained by an orthogonal transformation of the original p-axes, and only after the transformation, the number of orthogonal axes to be retained, that is m, is decided, taking the expected loss of information, among other things, into consideration.

Factor analysis (FA) has aims similar to those of PCA. There are, however, certain important differences between the two. In the case of PCA, there are no previous assumptions concerning the number and character of the principal components or factors to be extracted, whereas in the case of FA the following must be assumed:

(1) The number of common factors to be considered;
(2) The extent of contribution of each variable to the common factors.

The fundamental model of FA is expressed as follows:

$$x_i = a_{i1}f_1 + a_{i2}f_2 + \ldots + a_{ik}f_k + \ldots + a_{im}f_m + e_i \qquad (1)$$

where x_i is the standardized ith variable $(i = 1, 2, \ldots, p)$; f_k is the score of the kth common factor for each sample $(k = 1, 2, \ldots, m)$; a_{ik} is the factor loading for the ith variable and kth common factor; and e_i is error or the unique factor of the ith variable that is not explained by the m factors. The unique factor e_i is assumed to be orthogonal to all the common factors and to other unique factors associated with the other variables in the set. Therefore, if there is any correlation between the two variables i and j, it is assumed to be because of the common factors.

In this preceding model, the communality h_i^2 is defined as follows:

$$h_i^2 = a_{i1}^2 + a_{i2}^2 + \ldots + a_{im}^2 \qquad (2)$$

The communality, together with δ_i^2, or variance of the unique factor e_i, corresponds to the variance of the ith variable x_i or $V\{x_i\}$, thus

$$V\{x_i\} = h_i^2 + \delta_i^2 \tag{3}$$

Here, as x_i is standardized, $V\{x_i\} = 1$.

Another feature of FA is that the estimated factor can be rotated freely to make the interpretation of the factor easier. This is possible because of what is called indeterminacy of factor axes. If the factors, after rotation, are interpretable, computation of factor scores follows, based on the least square estimation.

The actual procedure to be used in this study consists of the following steps:

(1) Determination of the number of factors to be considered by means of PCA. A characteristic equation, in which $\boldsymbol{R}$ is the correlation matrix of the variables used;

$$|\boldsymbol{R} - \kappa\boldsymbol{I}| = 0 \tag{4}$$

is solved and the principal components having eigenvalues (κ) larger than 1 are considered.

(2) Preliminary communalities are computed as the squared multiple correlation between a variable and the rest of the variables in the set. In matrix form it is expressed as

$$\boldsymbol{H} = \boldsymbol{I} - \{\mathrm{diag} \cdot (\boldsymbol{R}^{-1})\}^{-1} \tag{5}$$

where $\boldsymbol{H}$ is a diagonal matrix with p-elements of h_i^2 ($i = 1, 2, \ldots, p$), and $\boldsymbol{R}^{-1}$ is the inverse of the correlation matrix. Since communality is defined as the proportion of a variable sharing something in common with variables in the set, the communality of a variable cannot be smaller than the squared multiple correlation.

(3) Principal factor analysis; starting from the reduced correlation matrix, the following characteristic equation is solved:

$$|(\boldsymbol{R} - \boldsymbol{I} + \boldsymbol{H}) - \lambda\boldsymbol{I}| = 0 \tag{6}$$

The predetermined (step 1) number of factors is extracted, and from eigenvalues and eigenvectors new communality estimates are obtained. The diagonal elements are then replaced with these new communalities. The same process is continued until differences in the two successive communality estimates become negligible.

(4) The factor loadings are computed from the final solution of eigenvalues and eigenvectors, as follows:

$$a_{ik} = \lambda_k^{1/2} \cdot l_{ik} \tag{7}$$

where a_{ik} is as before, λ_k is the kth eigenvalue, and l_{ik} is the ith element of the kth eigenvector that corresponds to the kth eigenvalue.

(5) The obtained factor axes are orthogonally rotated by Kaiser's varimax method, which aims at Thurston's simple structure to facilitate interpretation of the factors. Interpretation is easier when some of the variables make a high contribution to a factor and the other variables make a negligible contribution or

none at all. In other words, elements of a factor-loading vector should approach either of the extremes, unity or zero. Kaiser's varimax method maximizes the variance of the square of the elements of a factor-loading vector to attain this goal.

(6) If the results of rotation are reasonably interpretable, then the factor scores for individual soils are computed by the following general formula:

$$f_k = b_{1k}x_1 + b_{2k}x_2 + \ldots + b_{ik}x_i \ldots + b_{pk}x_p \qquad (8)$$

where b_{ik} is the factor score coefficient for the ith variable and kth factor. In the least square estimation, the $(m \times p)$ matrix $\boldsymbol{B}$ (factor score coefficient matrix) can be computed by the formula

$$\boldsymbol{B} = \boldsymbol{A'R}^{-1} \qquad (9)$$

where $\boldsymbol{A'}$ is a transposed factor-loading matrix, and $\boldsymbol{R}^{-1}$ is as used previously.

The characters used in this study are the eleven variables listed in Table 7.1. The selection was made taking the following into consideration:

TABLE 7.1   LIST OF CHARACTERS USED FOR ANALYSIS

Character No.	Name	Brief Description
1	TC (Total Carbon)	as percent of air-dried soil, Tyurin's wet combustion method.
2	TN (Total Nitrogen)	as percent of air-dried soil, Kjeldahl digestion and steam distillation.
3	NH_3-N	in mg N/100g of air-dried soil, after incubation for 2 weeks at 40°C.
4	Bray-P	in mg P_2O_5/100g of air-dried soil, Bray-Kurtz No. 2 method.
5	Ex-K	in me/100g of air-dried soil, N NH_4-acetate extraction, flame photometry.
6	CEC	in me/100g of air-dried soil, buffered neutral N $CaCl_2$ medium.
7	Av-Si (Available Silica)	in mg SiO_2/100g of air-dried soil, pH 4 Acetic acid extraction at 40°C.
8	TP (Total Phosphorus)	in mg P_2O_5/100g of air-dried soil, either HF-H_2SO_4 or HNO_3-H_2SO_4 digestion.
9	HCl-P	in mg P_2O_5/100g of air-dried soil, $0.2N$ HCl extraction at 40°C for 5 hrs.
10	Sand	as percent of organic matter-free dried soil, sum of coarse and fine sands.
11	Ex-Ca+Mg	in me/100g of air-dried soil, N NaCl extraction, EDTA titration.

(1) Primary data directly obtained from analysis are preferred.

(2) One of the items of complementary data is omitted. In other words, one of the characters that are highly negatively correlated is omitted, leaving the character which is easier to interpret, relative to fertility. For example, clay and sand are found to be highly negatively correlated (Table 5.24 in chapter 5). High sand content in soils clearly has a negative effect on fertility, while high clay content does not always have a positive effect. Therefore, the clay data was omitted.

Since multivariate normal distribution of the variables is implicitly assumed in FA, logarithmic transformation was used to make the positively skewed variables (chapter 5) approach the normal distribution.

7.2 EXTRACTION OF FERTILITY COMPONENTS

In chapter 5 the results of correlation analysis of the soil data were presented. The main conclusions were as follows:

(1) Characters related to base status, texture, and clay mineralogy are mutually highly correlated.

(2) Characters related to organic matter status are mutually highly correlated but are correlated only slightly to insignificantly with other characters.

(3) The same is true for the characters related to phosphorus status.

The correlation matrix for the log-transformed eleven characters used is given in Table 7.2. The correlations among the variables are generally improved, but the preceding three points still appear to be valid.

Starting from the correlation matrix of Table 7.2, PCA was run to estimate the number of factors to be considered. The largest 5 eigenvalues of the correlation matrix and the cumulative percentages of the variance are:

Eigenvalue	Percent of Total Variance	Cumulative Percent of Total Variance
5.501	50.0	50.0
1.836	16.7	66.7
1.673	15.2	81.9
0.698	6.3	88.3
0.372	3.4	91.6

Since the variances of the first three factors are larger than 1, and together account for more than 80 percent of total variance, and the fourth factor accounts for considerably less variance than the third, we

TABLE 7.2   CORRELATION MATRIX OF ELEVEN LOG-TRANSFORMED CHARACTERS

	(TC)	(TN)	(NH$_3$-N)	(TP)	(Bray-P)	(HCl-P)	(Ex-Ca+Mg)	(Ex-K)	(CEC)	(Av-Si)	(Sand)
(TC)	1.000	0.944	0.658	0.550	0.244	0.204	0.345	0.465	0.555	0.349	−0.442
(TN)	0.944	1.000	0.670	0.558	0.259	0.209	0.245	0.417	0.445	0.264	−0.404
(NH$_3$-N)	0.658	0.670	1.000	0.391	0.095	0.082	0.142	0.255	0.267	0.241	−0.215
(TP)	0.550	0.558	0.391	1.000	0.576	0.563	0.527	0.650	0.603	0.605	−0.274
(Bray-P)	0.244	0.259	0.095	0.576	1.000	0.859	0.217	0.405	0.222	0.260	−0.014
(HCl-P)	0.204	0.209	0.082	0.563	0.859	1.000	0.330	0.398	0.284	0.327	−0.022
(Ex-Ca+Mg)	0.345	0.245	0.142	0.527	0.217	0.330	1.000	0.775	0.912	0.809	−0.473
(Ex-K)	0.465	0.417	0.255	0.650	0.405	0.398	0.775	1.000	0.807	0.710	−0.509
(CEC)	0.555	0.445	0.267	0.603	0.222	0.284	0.912	0.807	1.000	0.812	−0.561
(Av-Si)	0.349	0.264	0.241	0.605	0.260	0.327	0.809	0.710	0.812	1.000	−0.310
(Sand)	−0.442	−0.404	−0.215	−0.274	−0.014	−0.022	−0.473	−0.509	−0.561	−0.310	1.000

NOTE: Character symbols in parentheses denote log-transformed characters.

decided to use only the first three factors in the next steps of factor analysis.

Using squared multiple correlation coefficients between a specific variable and the others in the set as the preliminary communality estimate, principal factoring with iteration was performed. After sixteen iterations, the final factor loadings and communalities were obtained, as shown in Table 7.3. The first factor is moderately to highly correlated with all the variables, while the second and the third, mainly with characters related to organic matter status and available phosphorus status, respectively.

The communality figures indicate that NH_3-N and Sand are only partially represented by the three factors. In other words, they have larger proportions of the unique factor not shared by the other variables.

The results of the varimax rotation are given in Table 7.4. Here interpretation of the factors is much easier. The first factor is highly correlated only with the characters related to base status and parent material, such as CEC, exchangeable cations, available silica, total phosphorus and sand, with the latter being opposite in sign to the rest. Thus, the first factor may be termed inherent potentiality (IP), and is determined primarily by the nature and amount of clay and base status.

The second factor is related to TC, TN, and NH_3-N. Moderately high loadings on TP and Sand are interpretable in terms of organic phosphorus and textural control on organic matter accumulation (see chapter 5), respectively. Therefore, the second factor may be termed organic matter and nitrogen status (OM).

The third factor can clearly be interpreted as available phosphorus

TABLE 7.3   FACTOR LOADING MATRIX AND COMMUNALITY FOR THREE PRINCIPAL FACTORS

	FACTOR 1	FACTOR 2	FACTOR 3	COMMUNALITY
(TC)	−0.744	−0.582	0.194	0.930
(TN)	−0.699	−0.635	0.304	0.984
(NH_3-N)	−0.455	−0.474	0.172	0.462
(TP)	−0.787	0.096	0.220	0.676
(Bray-P)	−0.502	0.464	0.664	0.908
(HCl-P)	−0.514	0.503	0.542	0.811
(Ex-Ca + Mg)	−0.889	0.083	−0.409	0.964
(Ex-K)	−0.794	0.282	−0.455	0.916
(CEC)	−0.833	0.173	−0.179	0.756
(Av-Si)	−0.746	0.232	−0.292	0.696
(Sand)	0.499	0.179	0.259	0.348

TABLE 7.4   TERMINAL FACTOR LOADING MATRIX FOR THREE
FACTORS AFTER VARIMAX ROTATION

	FACTOR 1	FACTOR 2	FACTOR 3
(TC)	0.288	0.913	0.118
(TN)	0.173	0.965	0.151
(NH$_3$-N)	0.113	0.668	0.046
(TP)	0.478	0.407	0.531
(Bray-P)	0.089	0.104	0.943
(HCl-P)	0.182	0.042	0.881
(Ex-Ca + Mg)	0.936	0.276	0.108
(Ex-K)	0.944	0.053	0.147
(CEC)	0.777	0.246	0.303
(Av-Si)	0.796	0.118	0.218
(Sand)	−0.486	−0.323	0.086

status (AP). Factor loading on TP is much less than on Bray-P and HCl-P. Contribution of other variables to this factor is minor.

It was observed that these three mutually orthogonal factors are in accordance with the result of correlation analysis referred to earlier. This leads to the interesting and important inferences that soil fertility of tropical Asian paddy soils is composed of at least three major components, and that both organic matter status and available phosphorus status of these soils are independent of what we call inherent potentiality.

7.3 EVALUATION OF FERTILITY FOR SAMPLE SOILS

The factor scores were computed for individual soil samples so that quantitative evaluation of the three components of soil fertility may be made. The coefficient matrix for the score computation is given in Table 7.5. Since the data were log-transformed before the analysis, the same transformation is needed for score computation. Moreover, the transformed data must further be standardized using the mean and standard deviation vectors given in Table 7.6. The scores are the sum of the products of the coefficient and the transformed and standardized datum corresponding to the coefficient, as shown in equation (8).

The scores thus computed for the samples are standardized with a mean of zero and a variance of unity. Therefore, positive score values indicate above-average status with reference to the overall mean for the 410 sample soils, and negative values indicate below-average status. In Table 7.7 soils with the ten highest and the ten lowest scores for each factor are listed.

The distribution of the factor scores is given in Table 7.8. The

TABLE 7.5   FACTOR SCORE COEFFICIENT MATRIX FOR THREE FACTORS

	FACTOR 1	FACTOR 2	FACTOR 3
(TC)	−0.151	0.268	−0.078
(TN)	−0.147	0.839	0.010
(NH$_3$-N)	0.045	−0.012	0.008
(TP)	0.051	−0.025	0.084
(Bray-P)	−0.091	−0.101	0.701
(HCl-P)	−0.059	−0.033	0.278
(Ex-Ca + Mg)	0.757	−0.132	−0.214
(Ex-K)	0.306	−0.144	0.029
(CEC)	0.130	0.018	−0.026
(Av-Si)	−0.058	0.073	0.087
(Sand)	0.028	0.012	0.004

TABLE 7.6   MEANS AND STANDARD DEVIATIONS OF THE ELEVEN LOG-TRANSFORMED CHARACTERS FOR 410 SAMPLE SOILS

VARIABLE	MEAN	S. D.
(TC)	0.044	0.297
(TN)	−0.994	0.282
(NH$_3$-N)	0.731	0.425
(TP)	1.775	0.429
(Bray-P)	0.171	0.584
(HCl-P)	0.608	0.712
(Ex-Ca + Mg)	1.159	0.342
(Ex-K)	0.993	0.484
(CEC)	−0.623	0.449
(Av-Si)	1.195	0.515
(Sand)	1.314	0.544

inherent potentiality scores show a somewhat negatively skewed distribution. The ten highest scores are for grumusols or grumusolic alluvial soils, six from India and four from Indonesia. On the negative side, fourteen scores are less than −2, of which one is even less than −3. Twelve of the fourteen are from the Northeast Plateau region of Thailand and the other two are from Malaysia and India. These are, without exception, sandy soils derived from strongly weathered parent materials.

In the case of organic matter-nitrogen status, the distribution of the scores is somewhat positively skewed. Here thirteen soils have scores greater than 2, of which six have scores even greater than 3. These extraordinarily high scores are for swamp soils, ten from West Malaysia

TABLE 7.7   LIST OF SOILS WITH THE TEN HIGHEST AND THE TEN LOWEST SCORES FOR EACH OF THE THREE FACTORS

IP			OM			AP		
In	33	1.908	M	20	3.484	B	8	3.208
In	29	1.774	Sr	27	3.265	I	60	3.101
I	38	1.738	M	40	3.229	I	58	2.959
I	55	1.728	M	22	3.198	I	56	2.265
In	31	1.709	M	21	3.056	T	62	1.946
I	57	1.645	M	19	3.003	I	8	1.923
In	22	1.633	In	4	2.906	Ph	44	1.919
I	65	1.605	Bd	48	2.417	I	27	1.898
I	74	1.570	M	39	2.404	Bd	15	1.825
I	53	1.564	M	41	2.329	Ph	43	1.781
T	20	−2.179	I	33	−1.802	In	33	−1.867
T	11	−2.182	I	38	−1.816	Sr	5	−1.889
M	14	−2.190	I	27	−1.821	T	72	−1.899
T	25	−2.198	I	11	−1.940	T	3	−1.902
T	15	−2.247	T	27	−2.014	T	15	−1.921
I	27	−2.261	Sr	1	−2.026	Ca	3	−1.945
T	16	−2.290	T	30	−2.113	Ca	11	−1.962
T	13	−2.691	T	18	−2.116	Ca	10	−1.987
T	14	−2.787	I	50	−2.347	Ca	4	−2.004
T	17	−3.181	T	23	−2.699	Ca	2	−2.202

Bd—Bangladesh	B—Burma	Ca—Cambodia
I—India	In—Indonesia	M—Malaysia
Ph—Philippines	Sr—Sri Lanka	T—Thailand

TABLE 7.8   DISTRIBUTION OF THE THREE FACTOR SCORES AROUND THE OVERALL MEAN (OR ZERO) FOR 410 SAMPLE SOILS

	IP	OM	AP
> 3	0	6	2
3 ~ 2	0	7	2
2 ~ 1	72	47	67
1 ~ 0	151	141	131
0 ~ −1	120	150	148
−1 ~ −2	53	53	58
−2 ~ −3	13	6	2
< −3	1	0	0

and one each from the Wet Zone of Sri Lanka, Bangladesh, and Indonesia; this last is exceptional, being an ando soil of volcanic ash origin. Many of the low-scored soils are from India and Thailand where climatic control seems to be more dominant than textural control.

Available phosphorus status shows nearly normal distribution. Three of the four soils having scores greater than 2 are from Andhra Pradesh, India; and the fourth is an alluvial soil from Burma. The poorest soils in terms of available phosphorus are from Cambodia; in fact, five of the ten lowest-ranked soils are from Cambodia.

Although some reservation regarding the sampling procedure adopted in this study is necessary, a rough estimation of fertility status can be made for each country or region by calculating a mean score for the samples concerned. Table 7.9 shows the mean values of the three factors for each country. The smallest standard deviation is for the organic matter-nitrogen status of Indian paddy soils (0.62) and the largest figure is for inherent potentiality of Thai soils (1.20). Inherent potentiality of Cambodian soils, organic matter-nitrogen status of Sri Lanka soils, and available phosphorus status of Burmese and Philippine soils show as large variance (about 1.0) as the corresponding variance for the total of 410 samples.

Inherent potentiality is highest for the soils of Indonesia and the Philippines, followed by those of India. The first two countries are situated in a region influenced by volcanic activity and the parent material of the soil is continuously rejuvenated by fresh volcanic ejecta. India is located in semiarid to subhumid climatic regions, and the weathering process of the soil material has not been very intensive,

TABLE 7.9   MEANS AND STANDARD DEVIATIONS OF THE THREE
FACTOR SCORES FOR THE RESPECTIVE COUNTRIES

COUNTRY	NO. OF SAMPLES	IP		OM		AP	
		Mean	S.D.	Mean	S.D.	Mean	S.D.
Bangladesh	53	−0.438	0.708	0.176	0.704	0.459	0.800
Burma	16	0.118	0.710	−0.128	0.646	0.308	1.121
Cambodia	16	−0.231	0.990	−0.155	0.954	−1.277	0.841
India	73	0.449	0.837	−0.780	0.619	0.581	0.906
Indonesia	44	0.618	0.766	−0.014	0.746	0.031	0.734
Malaysia	41	−0.545	0.719	1.398	0.927	0.026	0.726
Philippines	54	0.618	0.703	0.337	0.673	0.022	1.041
Sri Lanka	33	−0.510	0.853	0.150	1.064	−0.110	0.650
Thailand	80	−0.364	1.195	−0.347	0.939	−0.641	0.769

especially in the basaltic rock area of the Deccan Plateau that constitutes the catchment of the Godavari-Krishna rivers.

The soils of Malaysia and Sri Lanka, which are situated in permanently humid to monsoonal climatic regions of the low latitudes, are among the poorest with respect to inherent potentiality. The soils of Bangladesh, Cambodia, and Thailand are mostly on the poorer side of the overall mean.

Organic matter-nitrogen status is by far the highest for Malaysian soils, with a mean as high as 1.40. The second highest is for the Philippines, with a mean of 0.34. Conversely Indian soils, with the lowest mean, -0.78, are the poorest. The soils of the other countries are more or less similar, with mean scores clustering around the overall mean.

Available phosphorus status is high for the soils of India and Bangladesh, while Cambodian soils are the poorest, and next higher are the Thai soils. The regionality observed in this property may be ascribed to the climate and the parent material, or in other words, the degree of weathering of the soil material.

Similar calculations were carried out for regions that can be defined more or less discretely with respect to climate, parent material, and area. The result is given in Table 7.10. There are some regions in which one or more of the soil fertility components is still highly variable. The Northeast Plateau of Thailand and the east coast of Malaysia have very low mean scores for inherent potentiality. A great difference between these two regions, however, is seen in the standard deviation figures. In the Northeast Plateau of Thailand there occur narrow strips of clayey, recent alluvial soils and patches of grumusols associated with basalt outcrops, while the greater part of the region is covered by sandy, severely weathered and leached soils that mostly score -1.5 to -2 in inherent potentiality. By contrast, the soils of the east coast of Malaysia are very homogeneous in their material nature, all having been derived from strongly leached, medium to coarse textured alluvia. A similar comparison may be made between the Wet and Intermediate Zones of Sri Lanka and the east coast of Malaysia, having comparable mean organic matter-nitrogen scores but greatly different standard deviations.

The highest scores for inherent potentiality and available phosphorus status are for the soils of Godavari-Krishna region, which, however, have the second lowest organic matter-nitrogen score. The soils of the Northeast Plateau region of Thailand are characterized by very low scores for all three fertility components. High organic matter-nitrogen scores are shared by the soils of both the east coast and the Kedah-Perlis regions of Malaysia. Low available phosphorus scores are

TABLE 7.10   MEANS AND STANDARD DEVIATIONS OF THE THREE FACTOR SCORES FOR SELECTED REGIONS

REGION	No. OF SAMPLES	IP		OM		AP	
		Mean	S.D.	Mean	S.D.	Mean	S.D.
Sri Lanka Wet & Interm. Zone	14	−1.07	0.64	0.84	1.10	−0.10	0.64
Sri Lanka Dry Zone	19	−0.10	0.76	−0.36	0.70	−0.12	0.67
Bangladesh Ganges	15	0.33	0.35	0.04	0.82	0.90	0.50
Bangladesh Madhupur-Barind	9	−0.75	0.60	0.35	0.64	−0.14	0.40
Bangladesh Marginal	16	−0.87	0.62	0.36	0.66	−0.06	0.77
Bangladesh Brahmaputra	13	−0.57	0.48	−0.01	0.64	1.01	0.66
W. Malaysia Kedah-Perlis	10	0.25	0.44	1.21	0.56	0.02	0.54
W. Malaysia East Coast	10	−1.23	0.20	0.78	0.59	−0.54	0.57
India Godavari-Krishna	10	1.38	0.29	−0.73	0.47	1.11	0.95
Thailand NE Plateau	32	−1.18	1.27	−1.14	0.77	−0.94	0.70
Thailand Intermontane Basin	4	−0.27	0.52	0.06	0.69	−0.72	0.80
Thailand Upper Central Plain	14	−0.04	0.76	−0.05	0.49	−0.31	0.69
Thailand Bangkok Plain	24	0.53	0.64	0.24	0.66	−0.51	0.86
Thailand South	6	−0.40	0.74	0.58	0.48	−0.29	0.39

a common feature of all regions of Thailand. The Kedah-Perlis Plain of Malaysia and the Ganges sediment region of Bangladesh have relatively well-balanced soils with respect to the three fertility components.

To effect a fertility classification, the whole range of computed scores was divided into classes, with arbitrary class limits of ± 0.25 and ± 0.84. The assumption underlying the selection of the limits is that, if the distribution of the scores is normal, five classes of almost equal size should occur. The potentiality of each fertility component at different class levels could be designated as follows:

Class No.	Class limits	Potentiality
I	> 0.84	very high
II	0.84 ~ 0.25	high
III	0.25 ~ −0.25	intermediate
IV	−0.25 ~ −0.84	low
V	−0.84 >	very low

The distribution of the samples from each country or region using this classification is illustrated in figure 7.1, which is a collection of histograms showing percentage frequency of samples in each of the five fertility classes. Each of the regions in Table 7.10 is also shown. By referring to the figure, one can understand better the contents of Tables 7.9 and 7.10 and the contribution of each region to the make up of the samples of the respective countries.

Although clay mineral composition and total chemical composition were not directly used in this study, they are well represented by inherent potentiality. This is clear from Table 7.11, which shows the mean contents of clay mineral species and selected elemental oxides for each of the five inherent potentiality classes. The F-value in the last column is the variance ratio to test whether there is a difference in the mean contents among the classes. All the F values are much greater than the criterion value of $F(\phi_1 = 4, \phi_2 = 405; \alpha = 0.05) = 2.4$, indicating that the difference in the means is highly significant. Of the variables listed in the table, 10 Å mineral and total potash content have a peculiar pattern with their maximum in the intermediate classes.

Inherent potentiality is thus a compound character closely related to soil material characteristics. Therefore it naturally follows that it should be correlated with the soil material classes set up in the preceding chapter. To check this the mean and standard deviation of the inherent potentiality scores of samples in each material class were calculated (Table 7.12). Analysis of variance produced an F-value of 64.0, which is again much higher than $F(\phi_1 = 9, \phi_2 = 400; \alpha = 0.05) = 1.9$. When the ten material classes were arranged according to decreasing mean inherent potentiality scores, the result was as follows: VI ⩾ I > IV ≫ III > X > V ⩾ VII > IX ⩾ II ≫ VIII. A percentage distribution among the five inherent potentiality classes of the samples belonging to each material class is shown in Figure 7.2. High frequencies of inherent potentiality class 1 and 2 in material classes VI, I, and IV and the extremely high frequency of inherent potentiality class 5 in material class VIII are obviously seen from this figure.

In other words the majority of the samples in each material class fall into a narrow range of inherent potentiality classes. This supports

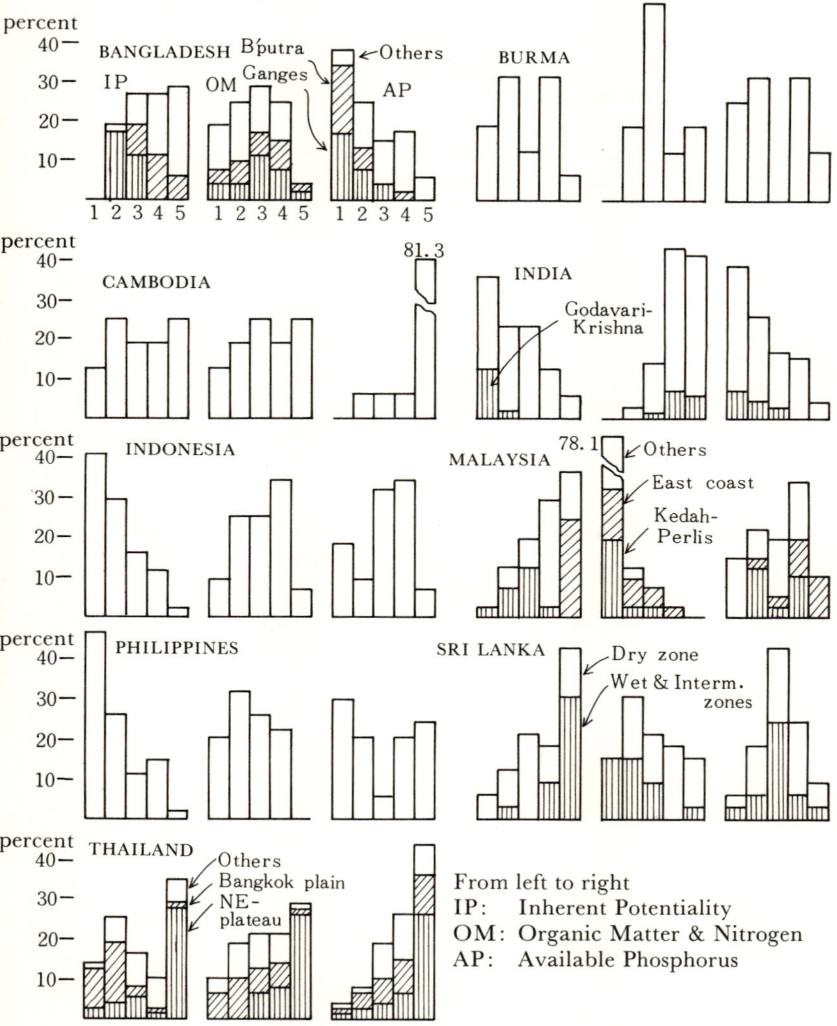

FIG. 7.1　PERCENTAGE DISTRIBUTION OF SAMPLES FROM EACH COUNTRY IN THE FIVE CLASSES OF EACH FERTILITY COMPONENT

the suggestion that the material classes be used for classifying soils at the "family" and, accordingly, "series" level, so that the latter will become more homogeneous with respect to soil capability.

In recent years many attempts have been made to derive a crop-yield prediction equation by means of multiple regression analysis, in

TABLE 7.11   MEAN CONTENTS OF CLAY MINERAL SPECIES AND SELECTED TOTAL ELEMENTAL OXIDES FOR THE SAMPLES IN EACH INHERENT POTENTIALITY CLASS

IP CLASS	1	2	3	4	5	F-VALUE
No. of Samples	(88)	(92)	(77)	(70)	(83)	
7 Å	27.84	40.27	44.03	52.71	67.83	51.34
10 Å	8.13	15.00	17.60	18.00	11.15	7.95
14 Å	64.03	44.73	38.38	25.50	19.82	78.09
SiO_2	63.86	66.80	71.61	74.94	84.89	73.46
Fe_2O_3	9.24	7.42	5.75	4.52	2.26	74.89
Al_2O_3	20.31	19.31	16.22	15.60	9.78	41.84
CaO	2.13	1.72	1.69	1.09	0.37	11.25
MgO	1.25	1.20	1.09	0.68	0.31	28.14
TiO_2	1.36	1.25	1.13	1.07	0.87	8.91
K_2O	1.50	2.00	2.28	1.90	1.53	5.90

TABLE 7.12   MEAN AND STANDARD DEVIATION OF INHERENT POTENTIALITY SCORES FOR THE SAMPLES BELONGING TO EACH SOIL MATERIAL CLASS

SOIL MATERIAL CLASS	INHERENT POTENTIALITY	
	Mean	S.D.
I	0.955	0.575
II	−0.405	0.676
III	0.159	0.656
IV	0.764	0.439
V	−0.188	0.766
VI	0.999	0.613
VII	−0.193	0.696
VIII	−1.613	0.694
IX	−0.391	0.636
X	−0.046	0.435

which soil characters are used as independent variables, together with such factors as climate and management that are thought to be relevant to crop-yield. Soil characters used in such attempts are often humus content, clay content, a certain nutrient content, and so forth. We believe that the three factor scores estimated in this study are most suitable to the purpose, because:

(1) the three factors are compound characters derived from many

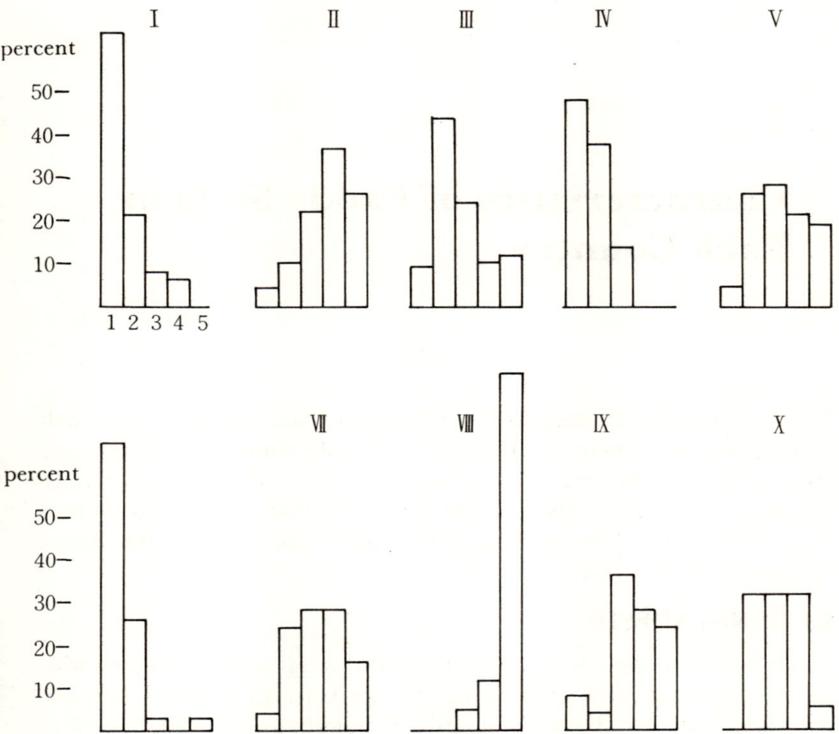

FIG. 7.2   PERCENTAGE DISTRIBUTION OF THE SAMPLES BELONGING
TO EACH OF THE TEN SOIL MATERIAL CLASSES (I ~ X) AMONG FIVE
INHERENT POTENTIALITY CLASSES

individual characters and represent the most important fertility components relevant to the yield, and

(2) these factors are mutually independent, so they best fit the multiple regression model.

Kyuma (1973b) showed that the three factors alone account for about 60 percent of the variation of paddy yield in a study with Malaysian data. In that particular study, inherent potentiality was shown to contribute most remarkably to paddy yield among the three factors.

8

Characteristics of Paddy Soils in Each Country

In the preceding chapters paddy soils in tropical Asia have been dealt with in a general manner. In most cases only country averages were given without discussing the regional differences within a country. In this chapter we try to give a more detailed account of the paddy soils in each country. For exact sampling locations, refer to the list of samples in Appendix 1.

8.1 BANGLADESH

Paddy soils in Bangladesh are distributed throughout the country with the exception of hilly areas along the northern and eastern borders, and the saline coastal areas, notably in Sunderbans. Most of the parent materials of the paddy soils are recent alluvial sediments deposited by the Brahmaputra (or Jamuna) and the Ganges (or Padma), and to a lesser extent by the Teesta, Meghna, and other minor rivers originating from the eastern Tertiary hills. There are two areas of Pleistocene terrace materials, the Madhupur Jungle and the Barind Tract. They are apparently more highly weathered and contain iron and manganese nodules. Some of the terrace soils have a peculiar morphological feature, the occurrence of a bleached horizon or whitish sand separations in the subsurface, on which Brinkman (1969) developed a discussion of "ferrolysis." Along the coast to the south of Chittagong, acid sulfate soils locally known as "kosh" soils occur in small areas (Islam, 1955). In Sunderbans, which is a wide coastal swamp forest area to the west of the present mouth of the Ganges-Brahmaputra, some potentially acid sulfate soils are known to occur (Brammer, 1970). But in this latter area, development of acidity would not be very severe because of the free lime contained in the Gangetic sediments.

No large areas remain for further expansion of rice lands unless coastal areas are effectively protected from saline water intrusion and from occasional flood tides caused by cyclones.

The sampling sites in Bangladesh are plotted in Figure 8.1. The samples may be grouped into the four according to the origin and the nature of parent materials (Brammer, undated map showing distribution of parent materials).

Gangetic alluvium   Bd-12 ~ 14, 35, 36, 44 ~ 53
Brahmaputra alluvium   Bd-1 ~ 3, 8, 10, 11, 15, 18, 19, 21, 22, 27, 29

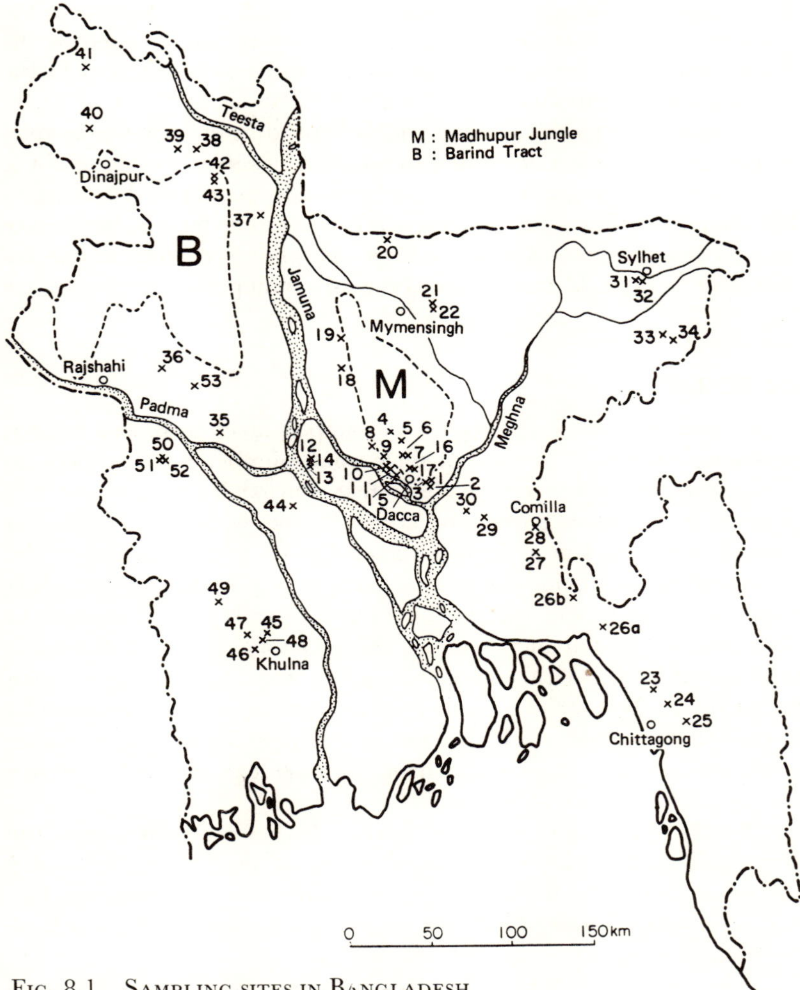

Fig. 8.1   Sampling sites in Bangladesh

Pleistocene terraces   Bd-4 ~ 7, 9, 16, 17, 42, 43
Marginal areas   Bd-20, 23 ~ 26, 28, 30 ~ 34, 37 ~ 41

As shown in Figure 8.2, A, soil materials are generally moderate in quality. Many of the soils on the Ganges and Brahmaputra sediments belong to soil material classes III, V, and VII, the last one being more silty and illitic in nature. The terrace materials and the materials in the piedmont of the Tertiary hills (included in marginal areas) are in class II, which is more depleted and siliceous.

Inherent potentiality grades 2 and 3 dominate in the Ganges and Brahmaputra sediment areas, while grades 4 and 5 are frequent in the terrace and piedmont areas. The northwestern region, covered by the Teesta sediments, is sandy in soil texture and this is reflected in the low inherent potentiality grade (see Fig. 8.2, B)

Organic matter status varies widely depending on local topography and no generalization according to region is possible (see Fig. 8.2, C).

Available phosphorus status is generally good for the Ganges-Brahmaputra sediment areas, whereas it tends to be poor in the terrace and piedmont areas. Relatively high phosphorus content is one re-markable feature of the recent Ganges-Brahmaputra sediments (see Fig. 8.2, D).

8.2 BURMA

As stated in chapter 4, the paddy soil samples taken in Burma are not satisfactory not only because of their small number but also because they represent paddy soils inadequately. Burma has two major rice growing areas, Lower Burma and Upper Burma, with entirely different climatic and physiographic conditions. Lower Burma comprises the Irrawaddi-Sittang delta and coastal areas facing the Bay of Bengal. It is humid to perhumid during the rainy season receiving more than 2000 mm of rainfall in 6 months (May-October). Upper Burma is represented by the area around Mandalay, characterized by a much drier climate. More than 55 percent of rice acreage is in the Irrawaddi-Sittang delta and about 18 percent is in Upper Burma. The coastal areas and the mountain states have 13.5 percent each of total rice acreage (Government of Burma, 1972).

Of the 16 samples taken in Burma, 6 (B-1 ~ 6) come from the Irrawaddi delta, 7 (B-7 ~ 13) from the area around Mandalay, and 3 (B-14 ~ 16) from the transition zone along the Sittang valley (see Fig. 8.3). B-7, taken in *kaing*-land or a sandbar in the Irrawaddi, and B-14 are nonpaddy soils.

The climatic difference between Lower and Upper Burma is reflected in the soil materials. As shown in Figure 8.4, A, the samples

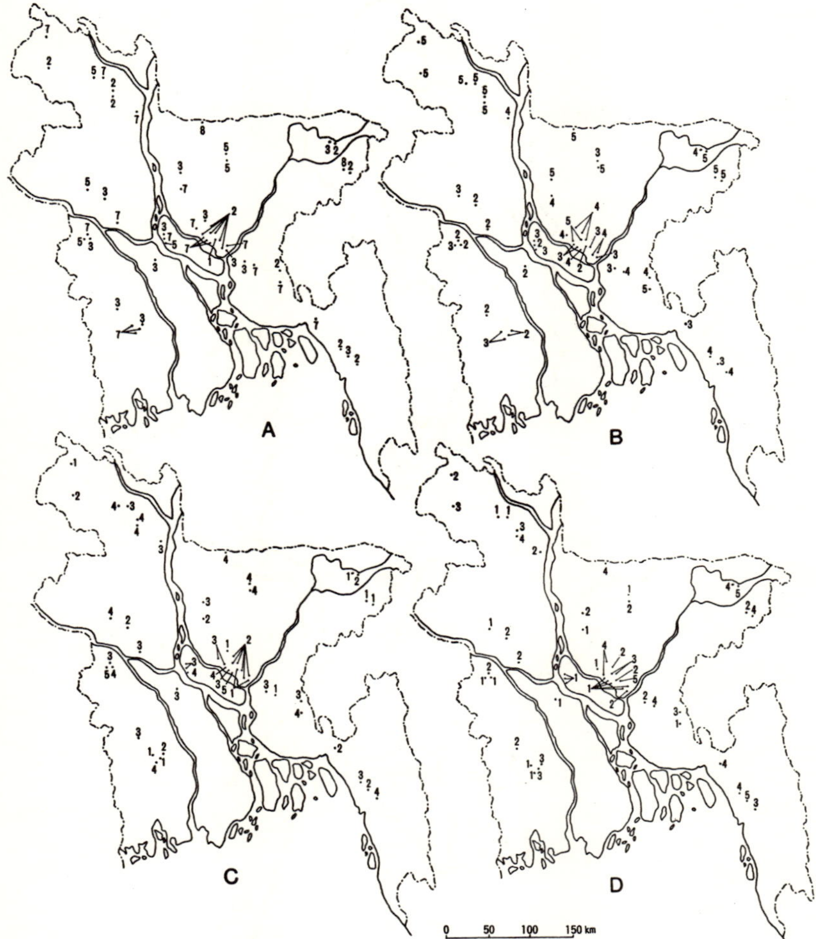

Fig. 8.2 Map of Bangladesh, showing distribution of samples in terms of: A, material class; B, inherent potentiality grade; C, organic matter and n grade; D, available p grade.

from the Irrawaddi delta are clayey but have moderate to low base status, class I, II, and III, while those from Upper Burma are characterized by moderate to high base status, classes IV, V, VII, and X.

Inherent potentiality is high for clayey soils in the delta and Upper Burma, but moderate to low for sandy nonpaddy soils and for those on older alluvia (B-6, 15, and 16) in the wetter climate areas (see Fig. 8.4, B).

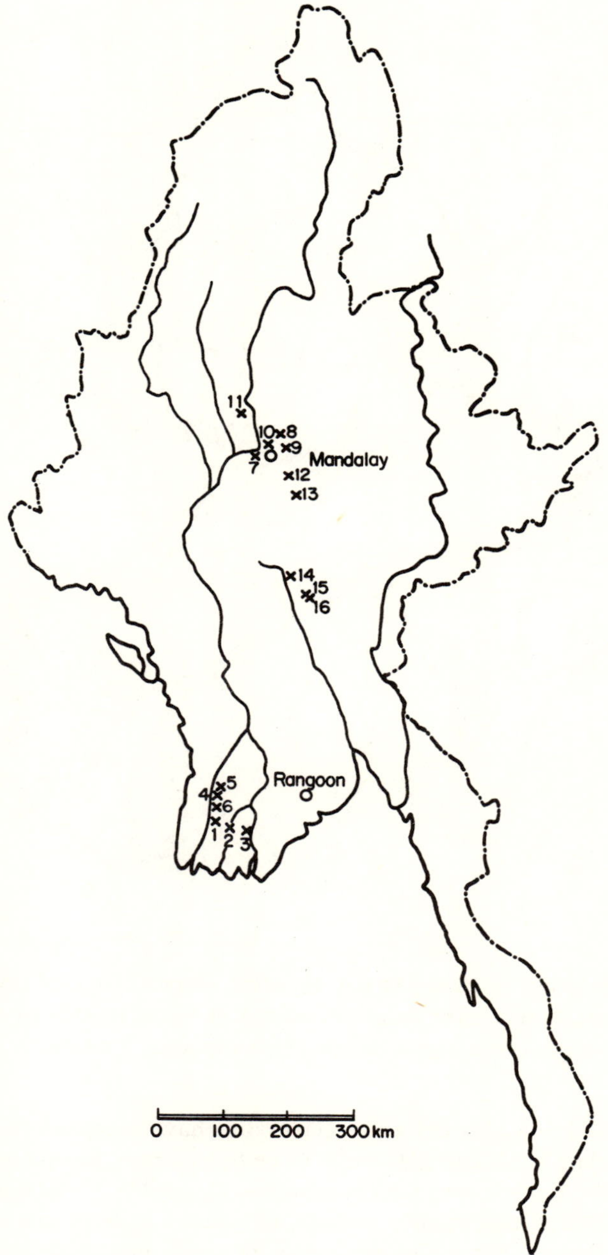

FIG. 8.3   SAMPLING SITES IN BURMA

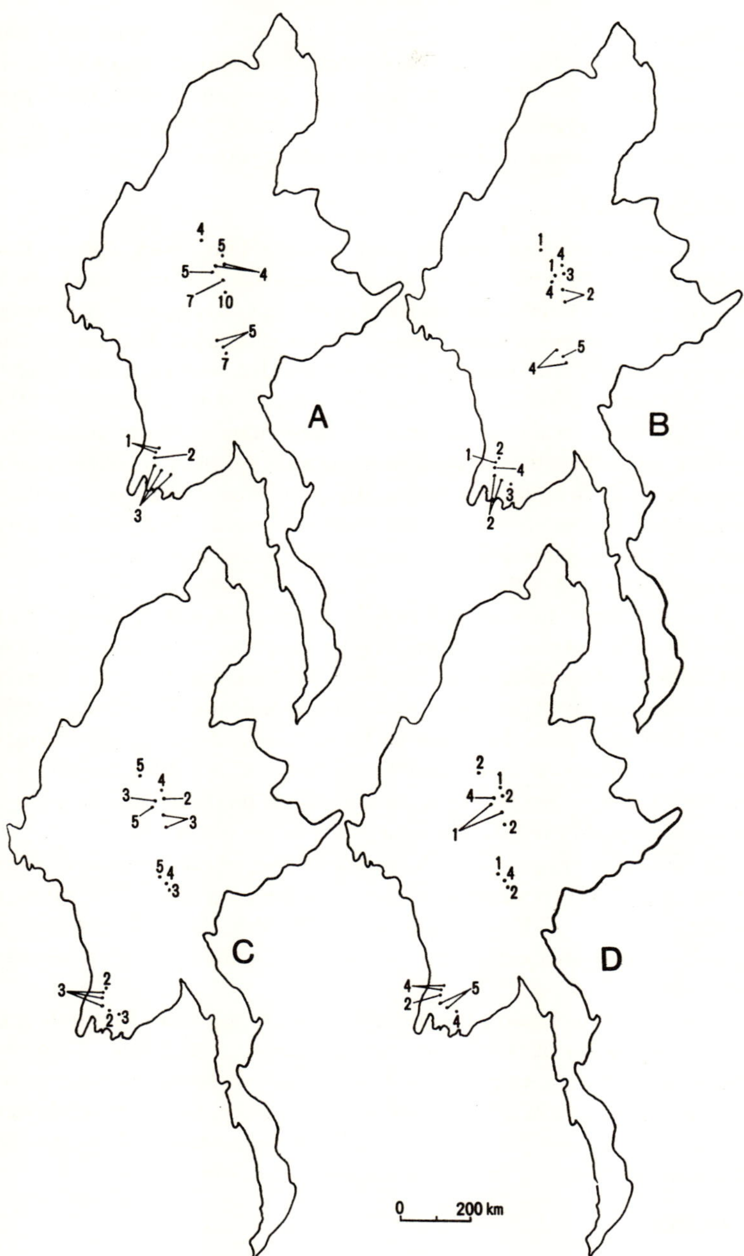

FIG. 8.4 MAP OF BURMA, SHOWING DISTRIBUTION OF SAMPLES IN TERMS OF: A, MATERIAL CLASS; B, INHERENT POTENTIALITY GRADE; C, ORGANIC MATTER AND N GRADE; D, AVAILABLE P GRADE.

Organic matter status is poorer for Upper Burma soils in a drier climate and richer for wetter deltaic soils. Phosphorus status is, on the contrary, poorer for the deltaic soils and generally better for Upper Burma soils (see Figs. 8.4, C and 8.4, D). Because of the inadequacy of the samples we cannot make further generalizations.

8.3 CAMBODIA

The Mekong floodplain and the area around Tonle Sap (Grand Lac) are the two areas for rice growing in Cambodia. The area downstream of Phnom Penh is part of the Mekong delta, but only a small percentage is utilized for rice cultivation because of the difficulty of water control. Physiographically, the rice lands are subdivided into two, one covered with relatively rich recent river and lacustrine sediments and the other with depleted terrace sediments derived mainly from weathered Mesozoic rocks. Basalt and limestone outcrops supply part of the parent materials, and produce exceptionally rich, for Cambodia, soils of grumusol nature among generally very poor soils. But rice is rarely grown on these better soils, which mainly support such commercial crops as rubber and coffee.

The sampling sites are plotted in Figure 8.5. Ca-1, 2, and 7 are derived from lacustrine alluvium. Ca-10 is on recent riverine sediment, and Ca-16 is taken from a low levee of the Mekong. Ca-13 is from an area delineated as Alumisol (acid sulfate soil) on Crocker's (1962) 1:1,000,000 soil map. It seems to be on the margin of the Plain of Reeds, a well-known acid sulfate soil area of the Mekong delta of Vietnam. Ca-6 is exceptional, being on a terraced basalt plateau developed as paddy fields. The others are all on old terraces.

As shown in Figure 8.6, A, many of the terrace soils are in soil material class XIII, poorest of all the soil material groups. The inherent potentiality grade of these soils is 5 (see Fig. 8.6, B). Only the soils on recent lacustrine sediments and the basalt plateau have high inherent potentiality.

Organic matter status is generally poor, as shown in Figure 8.6, C, with the exception of the lacustrine soils and the Alumisol. Available phosphorus status is very poor (see Fig. 8.6, D), with only one exception, the Mekong levee soil. All the fertility characteristics of this Mekong levee soil are similar to those of soils in a comparable physiographic position in the Mekong delta of Vietnam (see 8.10).

8.4 INDIA

Only the eastern states of India, with the exception of Assam, were surveyed. The samples collected represent roughly two-thirds of total

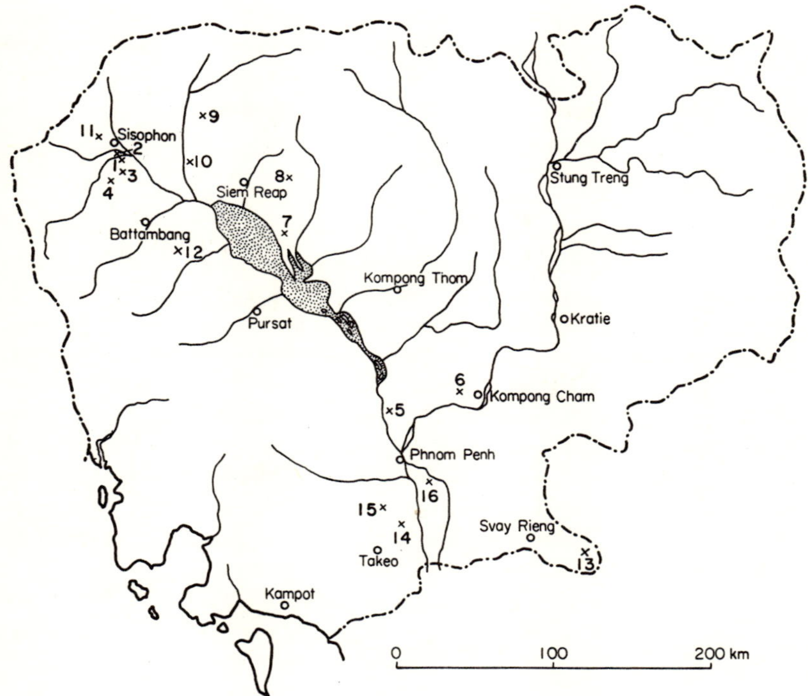

FIG. 8.5   SAMPLING SITES IN CAMBODIA

rice-growing areas. Variations in climate and parent material are so great that a simple generalization is not possible. Rice lands may be categorized into the following:

(1) Gangetic floodplain and terraces
(2) Deltas at the mouths of big rivers—Ganges, Mahanadi, Godavari-Krishna, and Cauvery
(3) Coastal lowlands, and
(4) Local alluvia in the interior of the Deccan peninsula.

Figure 8.7 shows the sampling sites. I-1 ∼ 21 are in the Ganges basin. Of these I-5 and I-16 are exceptionally rich in free lime, which derives from the sediments of the Gandak River, a tributary of the Ganges. I-7 is a grumusol locally called *karail* soil. I-4 was sampled in an area called *terai*, a piedmont plain of the foothills of the Himalayas. I-17 ∼ 19 are recent Kosi River sediments.

Soils in the second category include I-22 ∼ 26 from the Ganges-Hooghly delta, I-28 ∼ 31 from the Mahanadi delta, at the apex of which

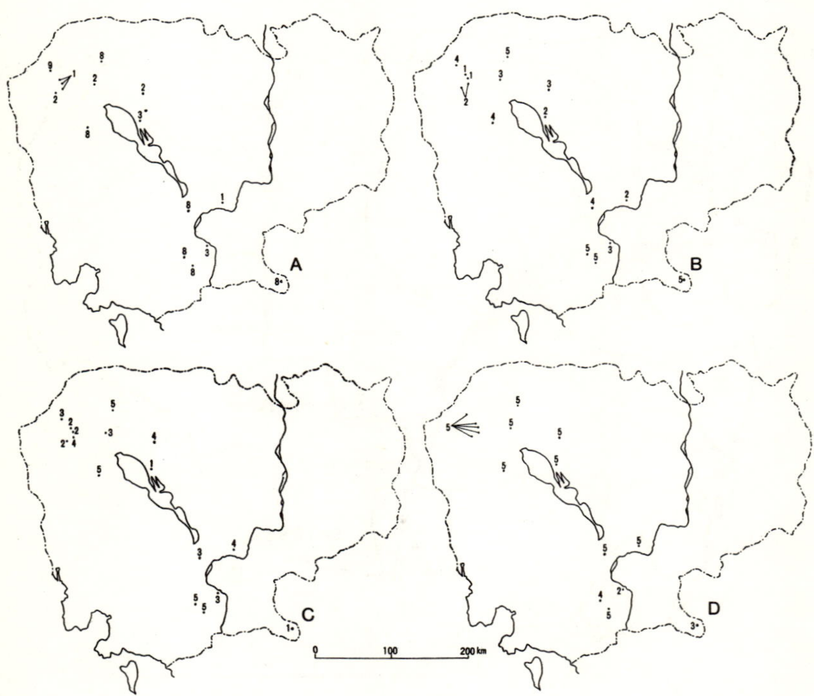

FIG. 8.6   MAP OF CAMBODIA, SHOWING DISTRIBUTION OF SAMPLES IN
TERMS OF: A, MATERIAL CLASS; B, INHERENT POTENTIALITY GRADE;
C, ORGANIC MATTER AND N GRADE; D, AVAILABLE P GRADE.

Cuttack is located, I-51 ~ 57 and I-65 and 66 from the Godavari-
Krishna delta, and I-74 from the Cauvery delta. I-52 is on a terrace
remnant within the Godavari delta.

Soils in the third category, coastal lowland soils, comprise I-27,
38 ~ 40, 43 ~ 50, 58, 67 ~ 73, and 75. I-27 was taken from an area in
which many abandoned laterite quarries are to be found. The soil
materials of I-43, 67, 70 ~ 72 are also thought to be derived from nearby
lateritized areas. I-58 and I-75 are on terraces at the periphery of deltas.
I-48 ~ 50 and I-68 and 69 are on coastal sand ridges.

I-32 and 33, I-34 ~ 36, I-41 and 42 are derived from local alluvia
in the interior valleys to the west of the Eastern Ghats. I-34 ~ 36 are
from a catena; I-34 is the lowest-lying sample, a lime-containing
grumusol locally called *kanhar*, while I-36 and 35 are on successively
higher positions and have no free lime in the upper part of the profile.
I-59 ~ 63 were sampled on the way to and around Hyderabad. Except

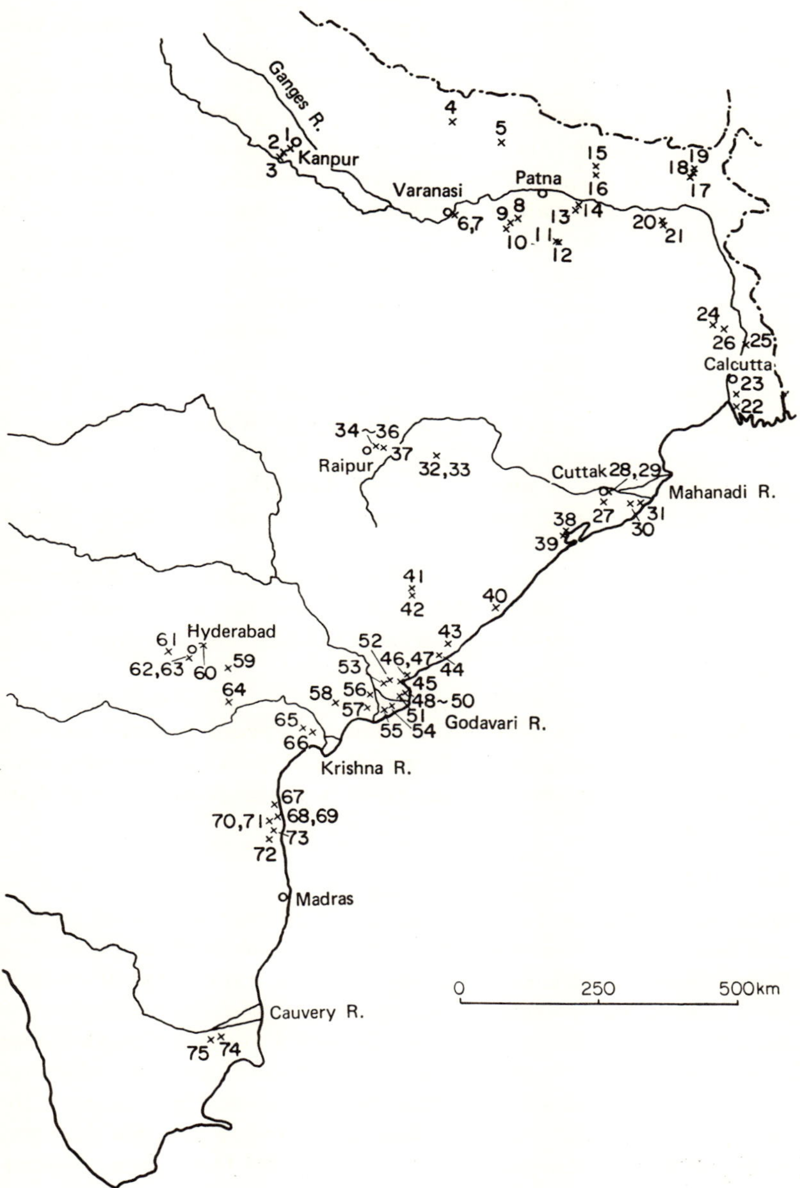

FIG. 8.7 SAMPLING SITES IN INDIA

for I-63, which is at the foot of a granite hill, all the rest contain free lime. I-62 is a typical grumusol on recent alluvium.

Soil materials are varied as seen in Figure 8.8, A. However, there seem to be some regionality in the occurrence of different soil material classes. For example, class VII occurs exclusively in Gangetic alluvia, which gave rise to many class VII soils in Bangladesh also. Class VI is confined to interior valleys, the Hyderabad area, and the Godavari-Krishna delta. Most of the class II samples occur in the Ganges terraces. Similarly distribution of class I and class IX are confined to a relatively narrow area around the mouths of the Godavari and Krishna rivers. Two class X samples are on lime-rich soils of the Gandak sediments.

Inherent potentiality, as shown in Figure 8.8, B, is high for soils on deltaic sediments and basin soils on local alluvia. In fact, the soils of the Godavari-Krishna delta have the highest inherent potentiality of all the geographical regions (see chapter 7). Soils with low inherent potentiality grades, 4 and 5, are rare, occurring only on the sandy alluvia of the Ganges tributaries and on strongly weathered terrace sediments.

Organic matter status is poor for most soils, probably a reflection of the drier climate (see Fig. 8.8, C). Only two soils, in basins of rolling terrain in the interior of the Deccan Peninsula, have organic matter status grade 2.

On the contrary, available phosphorus status is generally moderate to high, as seen in Figure 8.8, D. The only exceptions are two soils derived from the exceptionally lime-rich sandy Gandak sediments. They have available phosphorus status grade 5, probably due to phosphate fixation by lime.

8.5 INDONESIA

Only the island of Java was covered, but 70 percent of the total population live on Java and more than 60 percent of total paddy land in Indonesia is found there. The most prominent feature of paddy soils in Java is their parent material of volcanic origin. Physiographically, paddy soils occur either on river and coastal alluvia or on the lower slopes of volcanoes of pyroclastic materials. According to the Indonesian system of soil classification (Soil Research Institute, 1960), many of the latter soils are latosols, mediterranean soils, and regosols, but they may be grouped as Eutropepts and Dystropepts of the U.S. system, depending on their base status. Reflecting the variation in climate, both vertical and horizontal, and also slight differences in material, the soils in West Java are generally more depleted than those in Central and East Java.

The sampling sites in Java are shown in Figure 8.9. The samples

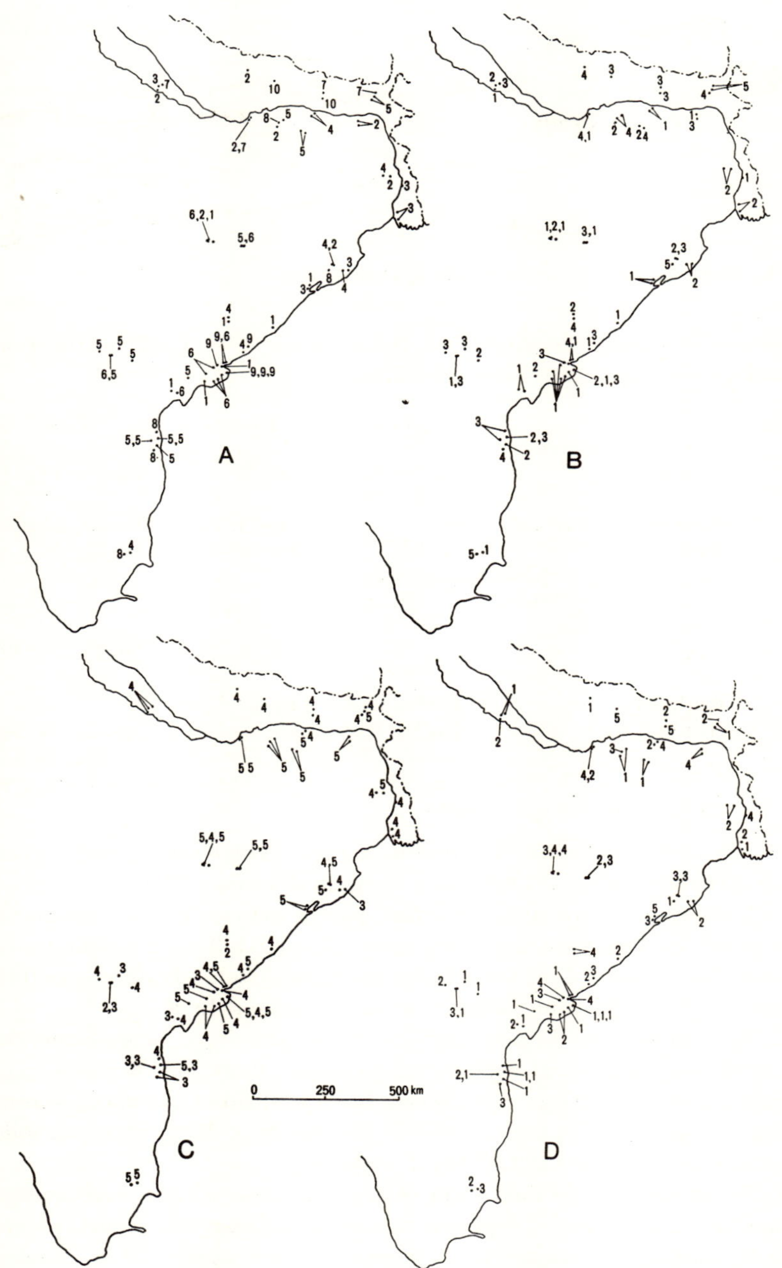

FIG. 8.8   MAP OF INDIA, SHOWING DISTRIBUTION OF SAMPLES IN TERMS OF: A, MATERIAL CLASS; B, INHERENT POTENTIALITY GRADE; C, ORGANIC MATTER AND N GRADE; D, AVAILABLE P GRADE.

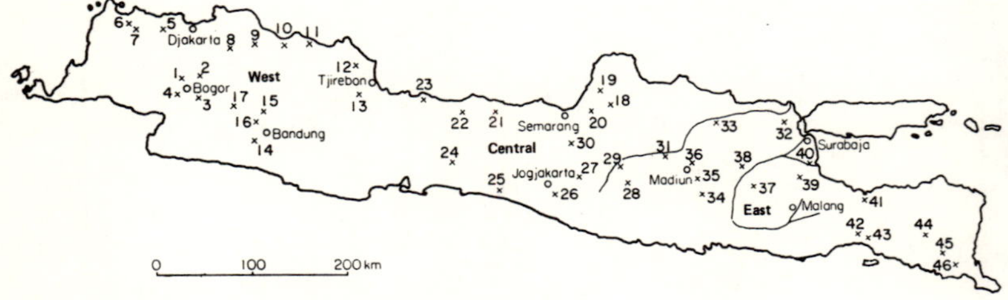

Fig. 8.9   Sampling sites in Java, Indonesia

may be grouped, according to the Indonesian system of soil classification and by geographical regions, as follows:

	West	*Central*	*East*
Latosols	In-1 ~ 3, 5, 13, 16	In-28	In-34, 44
Mediterranean soils	—	In-19, 29	In-35, 41
Regurs	In-17	In-20	In-31, 46
Gray hydromorphic and Red-yellow podsolic soils	In-6 ~ 8, 10	—	—
Regosols	—	In-26, 27	In-37, 39, 42, 45
Andosols	In-4, (15)*	(In-30)*	
Alluvial soils	In-9, 11, 12, 14	In-18, 21 ~ 25	In-32, 33, 36, 38, 40, 43

*Two andosols, In-15 and 30, are nonpaddy soils and are not included in the discussion.

Brief descriptions of the conditions under which these groups occur are given according to Dudal and Soepraptohardjo (1960).

1. Latosols occur mainly on volcanic parent materials. They are found from sea level to elevations of 900 m on rolling to hilly and mountainous land. The climate of the latosol area is wet, with annual rainfall ranging from 2500 to 7000 mm. Latosols occur throughout Java.

2. Red-yellow and brown mediterranean soils have been found on limestones, old alluvial deposits, and volcanic materials. Relief is level from sea level to 400 m. The climate has a well-pronounced dry season and annual rainfall ranges from 800 to 2500 mm. Mediterranean soils occur extensively in Central and East Java.

3. Regur soils are formed from marls, calcareous shales, argillaceous limestones, old alluvial deposits, and volcanic materials. Relief is level to undulating, elevation ranging from sea level to about 200 m. Annual rainfall ranges from 800 to 2000 mm. Regur soils occur over large areas

in Central and East Java. They were formerly called Margalite soils, in Indonesia.

4. Gray hydromorphic soils include planosols and low humic gley soils. They occur along the north coast of Java in association with red-yellow podzolic soils.

5. Regosols are soils from deep, unconsolidated material, other than alluvium, showing weak or no profile differentiation. Regosols on volcanic ash and tuffs are widely distributed in Java.

6. Andosols in Java are found on unconsolidated volcanic materials. They occur on undulating to mountainous land from sea level up to elevations of about 3000 m, the majority at higher elevations. They are mostly found in cool and high rainfall areas.

7. Alluvial soils occur extansively along the north coast of Java. Wide alluvial plains are also found along important rivers, such as the Solo, Brantas, Tjimanuk, and Tjitaram.

As seen in Figure 8.10, A, soil material classes I and VI are dominant in Java. Class X is also found, corresponding to the occurrence of regosols on fresh volcanic sands. Two soils belonging to class II, a poor material, are gray hydromorphic soils on the northern coastal terraces of West Java.

Inherent potentiality is generally high, as shown in Figure 8.10, B. Poor soils with inherent potentiality of grade 4 or 5 are exclusively found in West Java and are either gray hydromorphic soils or latosols. An andosol is also grade 4, probably because of sandy texture and low base status.

Organic matter status is generally low for the soils in Central and East Java as a reflection of the drier climate (see Fig. 8.10, C). Soils in grade 1 for organic matter status are all on higher elevations in West Java, where the climate is humid throughout the year.

Available phosphorus status is quite variable (Fig. 8.10, D). Regosols on volcanic sand in Central and West Java and the soils on the Solo alluvia are grade 1, while alluvial soils and latosols around Mt. Lawu on the border between Central and East Java are very poor in available phosphorus status, that is, grade 5.

8.6 MALAYSIA

A detailed account of paddy soils in West Malaysia is given in our earlier treatise on *Lowland Rice Soils in Malaya* (Kawaguchi and Kyuma, 1969b). As stated in that book, the climate of West Malaysia is Köppen's tropical rainforest climate (Af) except in a small area in the northwestern part, that is the Kedah-Perlis plain. Geologically, granite is the most important rock species, occupying about half of the total land surface.

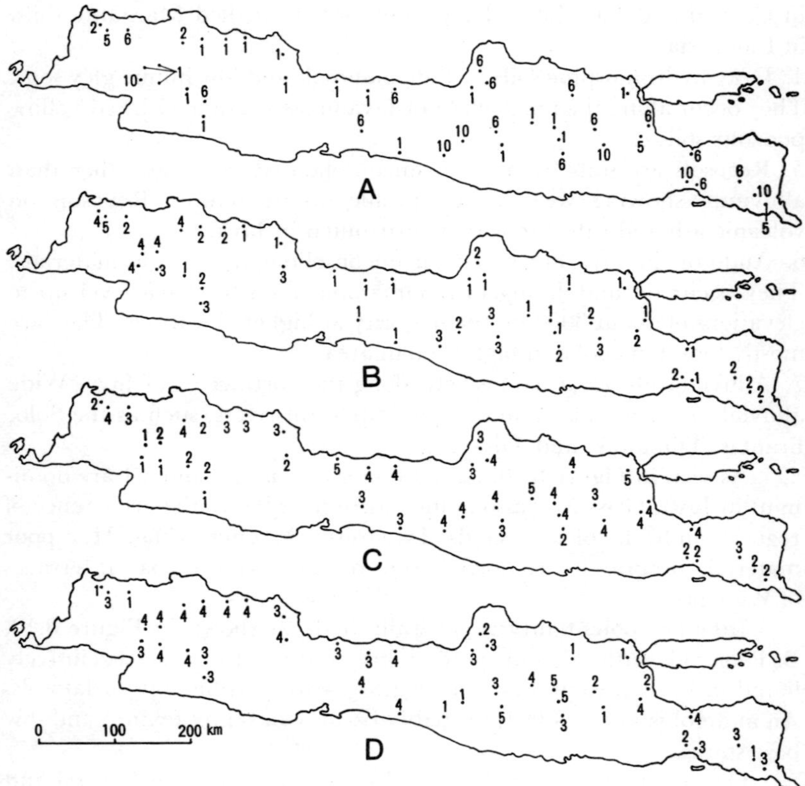

FIG. 8.10 MAP OF JAVA, INDONESIA, SHOWING DISTRIBUTION OF
SAMPLES IN TERMS OF: A, MATERIAL CLASS; B, INHERENT POTENTIALITY
GRADE; C, ORGANIC MATTER AND N GRADE; D, AVAILABLE P GRADE.

The sampling sites are plotted in Figure 8.11. Two broad subdivisions may be set up; soils on the west coast and those on the east
coast. Many paddy soils on the west coast occur in more or less swampy
conditions, reflecting the climate and the flat terrain. The soils of the
Krian area of Perak, M-17 ~ 21, are notable examples. Of the west
coast soils, those of the Kedah-Perlis plain (M-1 ~ 10) deserve special
attention. The climate of the area is characterized by a dominantly dry
period in January and February, and, therefore, swamp conditions are
less prevalent. The soils of this area have certain morphological simi-
larities to those of the Bangkok Plain of Thailand.

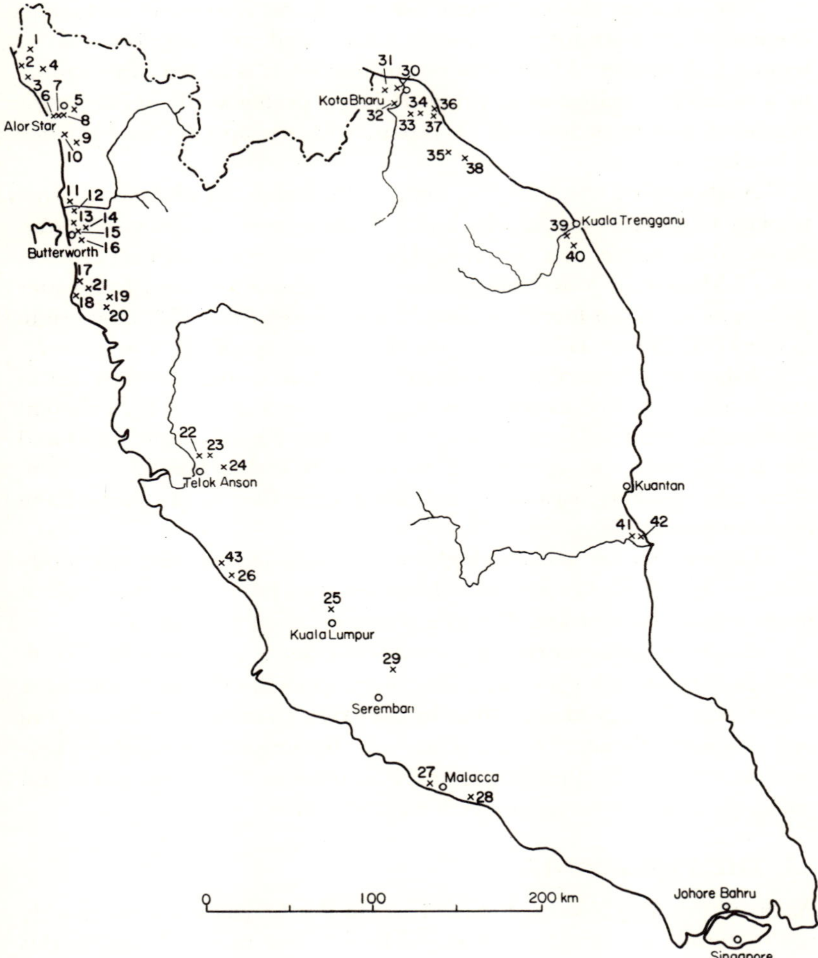

Fig. 8.11　Sampling sites in Malaysia

Many of the west coast soils originated from marine or brackish clay. Actually, most soils of the Kedah-Perlis plain, except M-1 and 5, and two of the Krian soils (M-17 and 18) are of this type. Two of the brackish soils, M-9 and 27, are active acid sulfate soils.

Although the soils of the west coast are generally clayey, M-14 and 24 are very sandy, because they are derived from alluvia of granite origin.

The soils on the east coast are mostly on freshwater sediments originated from granitic mountain ranges and are therefore coarser textured. A sample, M-36, was taken from what is called "bris sand," or a subrecent coastal sand ridge. It has a profile very similar to the degraded *akiochi* paddy soil of Japan, with an apparently bleached subsurface horizon.

As shown in Figure 8.12, A, many of the soil materials belong either to class II or III, both of which are characterized by low base status. As stated in chapter 5, a very low pH is the most remarkable characteristic of Malaysian soils. Soils derived from coarse-textured freshwater sediments are often found in class II and those of brackish clay origin in class III. Class VIII samples are all very sandy soils rich in quartz.

Inherent potentiality is generally low due to the low base status and kaolin-rich clay, as stated in chapter 5 (see Fig. 8.12, B). Inherent potentiality grades 1 and 2 occur only in the Kedah-Perlis plain and the Krian area and correspond to the soils of brackish clay origin. Generally speaking, east coast soils are poorer than west coast soils in inherent potentiality.

Organic matter status is high to very high, except for a few sandy soils (see Fig. 8.12, C). In fact, Malaysian soils have by far the highest mean organic matter and nitrogen score, as stated in chapter 7.

Available phosphorus status varies widely, as shown in Figure 8.12, D. This is partly a result of fertilizer application, which was more common in West Malaysia than in the other countries, at the time of our sampling. Generally east coast soils are lower in available phosphorus status, reflecting partly the low reserves of the material and partly the less intensive management in this area.

8.7 THE PHILIPPINES

As the Philippines is located in the Circum-Pacific volcanic zone, paddy soil materials are strongly influenced by volcanic materials and in this respect they are generally similar to those in Java. But as climatic conditions vary very greatly within the archipelago, soil variation is also great.

Figure 8.13 shows the sampling sites. The following six broad geographic regions were covered:

Luzon Central Plain	Ph-1 ~ 25
Cagayan valley	Ph-26 ~ 32
Bicol peninsula	Ph-50 ~ 54
Panay coastal plain	Ph-33 ~ 35
Southern Mindanao	Ph-36 ~ 45
Northern Leyte	Ph-46 ~ 49

FIG. 8.12   MAP OF MALAYSIA, SHOWING DISTRIBUTION OF SAMPLES IN TERMS OF: A, MATERIAL CLASS; B, INHERENT POTENTIALITY GRADE; C, ORGANIC MATTER AND N GRADE; D, AVAILABLE P GRADE.

The soils in the Luzon Central Plain are strongly affected by recent volcanic materials. Only in the eastern periphery are paddy soils developed on older materials, on dissected terraces, of which Ph-21 is an example. The region has a wet and dry monsoon climate, which seems to have led to the occurrence of grumusols and grumusolic alluvial soils on the volcanic materials.

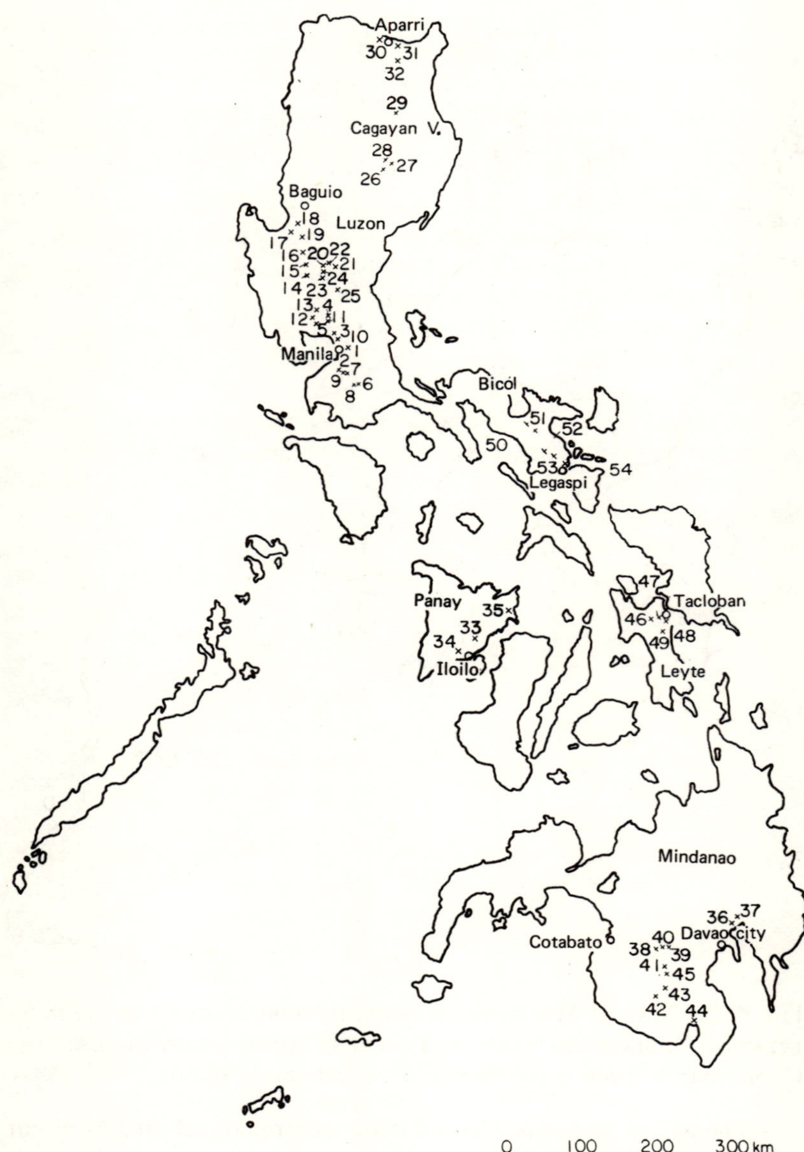

FIG. 8.13 SAMPLING SITES IN THE PHILIPPINES

Cagayan valley has a narrow strip of floodplain with relatively elevated terraces. Ph-27 was sampled on a well-weathered river terrace, which had many iron and manganese nodules. Ph-31, sampled near Aparri, shows many layers of buried A horizon, thus fitting the concept of Fluvent in the U.S. classification. Ph-30 is a brackish clay soil sampled on the northern coastal plain.

Bicol peninsula is another area of strong volcanic activity, as represented by the beautiful Mt. Mayon near Legaspi. All the samples are apparently influenced by volcanic materials. The climate is more humid here than in the Luzon Central Plain.

Panay coastal plain soils are also affected by volcanic materials, but one sample, Ph-35, taken on a terrace, is of strongly weathered material.

Of the Mindanao samples, Ph-36 and 37 are from the Davao area, the former being a brackish swamp soil. Ph-38 to 45 were sampled in Cotabato; Ph-39 was taken from an old terrace remnant with laterite block in the subsoil. The others are all derived from recent alluvia of volcanic material origin.

The Tacloban area of Leyte is perhumid throughout the year and often experiences strong typhoons. At the time of sampling, a problem for the people working in the paddy fields was the possibility of infection with schistosomiasis. Many of the low-lying paddy fields from which we took samples do not dry up at any time of the year.

The soil material classes are shown in Figure 8.14, A. Class I and IV are most frequent, both being the materials often encountered in volcanic regions. Soils on fresh volcanic sands without exception fall into class X. Three of the soils on old terraces (Ph-21, 27, 35, and 39) fall into class IX and one into class II, both classes are characterized by siliceousness and low base status.

Inherent potentiality is high for Luzon Central Plain soils, with the exception of a terrace soil, Ph-21 (see Fig. 8.14, B). The lowland soils of other regions have also high potential. The relatively low potential of the soils around Tacloban, Leyte, may be explained by stronger leaching and weathering of parent materials under perhumid conditions. Besides the terrace soils, many of the soils on recent volcanic sands are also graded 4 and 5 because of their coarse texture.

Organic matter status clearly reflects rainfall conditions (see Fig. 8.14 C). In Bicol and Leyte where the climate is wetter, organic matter status grade 1 occurs frequently, whereas grades 3 and 4 dominate in the Luzon Central Plain and Panay where the dry season is pronounced. Mindanao seems to be intermediate in this respect.

Available phosphorus status is highest in Cotabato, Mindanao, as

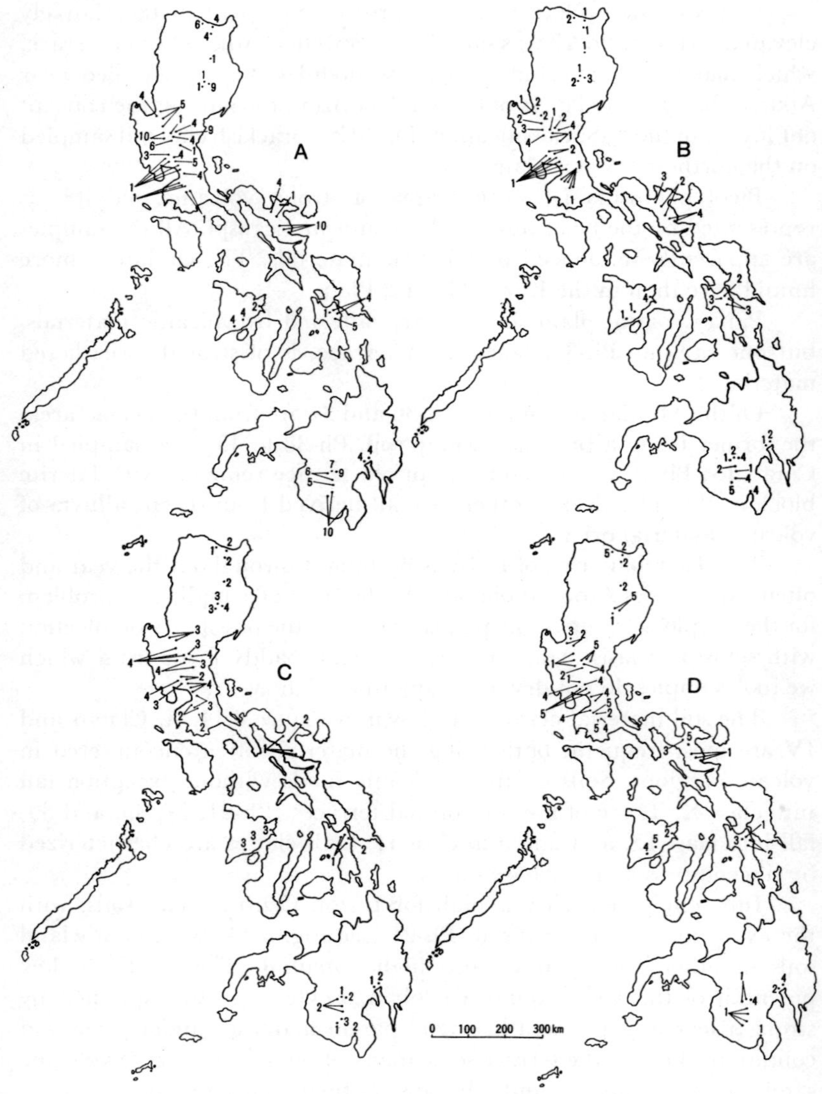

Fig. 8.14 Map of the Philippines, showing distribution of samples in terms of: A, material class; B, inherent potentiality grade; C, organic matter and N grade; D, available P grade.

seen in Fig. 8.14, D. Luzon Central Plain soils are not well endowed with this phosphorus, except for those on sandy volcanic material. In fact, soils on fresh volcanic sands all have an available phosphorus status of grade 1. We do not know why this high available phosphorus status cannot be retained by clayey materials of a similar nature in the Luzon Central Plain.

8.8 SRI LANKA

Paddy fields occur mostly on the coastal plain and on the lowest peneplain surface; the middle peneplain has a small area of rice land in narrow dissected valleys. Geologically 80 percent of the island consists of Paleozoic and Precambrian metamorphic rocks, of which garnet-sillimanite schists and gneisses are the most widespread, and therefore the most important sources of soil parent materials. Miocene limestones, occurring extensively in the northwestern part and in Jaffna are another important geologic formation (Cooray, 1967).

Climatic variation is great within the island. About a quarter of the total area, the southwestern part, is usually called the Wet Zone; and the greater part of the rest is called the Dry Zone, leaving a narrow strip between the two, the Intermediate Zone (see Fig. 8.15). Even the Dry Zone has 40–75 inches of rain annually, but the rainfall is unevenly distributed with the maximum during the northeast monsoon season, or Maha (December to February). Relatively small areas of the Dry Zone are used for cultivation during the southwest monsoon season, Yala, and they depend on tank irrigation, which is of ancient origin.

Sampling sites are plotted in Figure 8.15. Because there are few samples from the Intermediate Zone, these were combined with those from the Wet Zone for computing the regional means of fertility component scores (see chapter 7). Of the Dry Zone samples, Sr-1, 3, 5, 8, 9, 17, 30, 31, and 33 are from coastal plain; Sr-2 and 3 occur on low terraces and their material seems to have come from lateritized higher terraces further inland. Sr-5 and 8 are grumusolic alluvial soils, the materials of which were derived from the Miocene limestone area. Near Sr-5, gilgai microrelief was observed. The remaining Dry Zone samples were derived from local alluvia on the undulating to rolling terrain of the low and middle peneplains.

Since no wide coastal plains have developed along the Wet Zone coast, all the soils are considered to have been derived from local alluvia. A slight complication occurs in the case of Sr-28, where peaty organic matter accumulates in the surface and crude ambers were found in the organic layer.

As shown in Figure 8.16, A, soil material class IX predominates

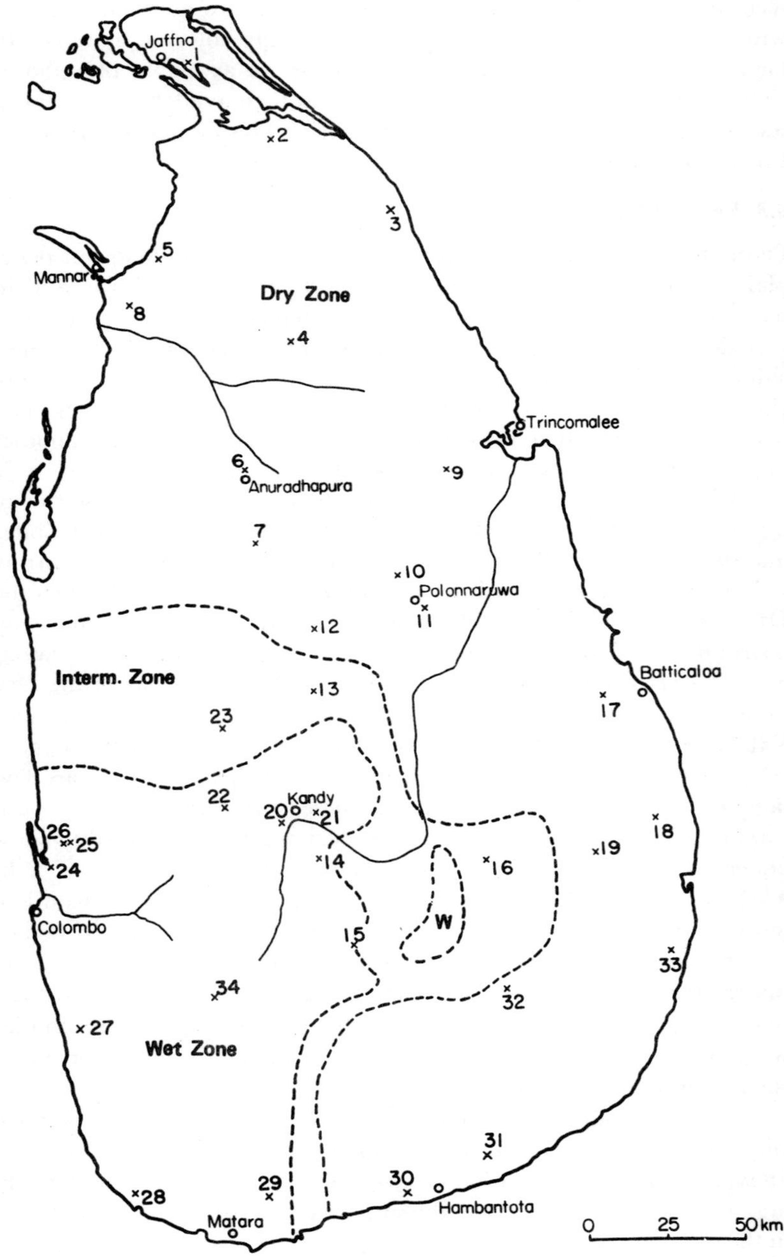

FIG. 8.15 SAMPLING SITES IN SRI LANKA

among the soils on the low peneplain. This class of material is characterized by a coarse texture and low base status, and in particular by an exceptionally low silt content. Fair numbers of classes V and VIII are seen, which are also characterized by coarse to very coarse texture. In the Dry Zone, heavy textured soils have either class I or VI material, while in the Wet Zone a heavy textured soil falls into class II, the siliceous and base poor group. Two class X soils seem to have been derived from charnockite, a dark-colored hyperthene containing metamorphosed rock that occurs as narrow bands regularly interbedded with quartzites, garnet-bearing gneisses, and so forth.

As shown in Figure 8.16, B, inherent potentiality of Wet and Intermediate Zone soils is very low with the exception of Sr-21, which corresponds to one of the soil material class X samples. The soils in the Dry Zone are usually a little better than the Wet Zone soils, because their base status is better, though the texture of both is coarse. Grumusolic soils in the limestone area, Sr-5 and 8, have high potential, with an inherent potentiality of grade 1.

Organic matter and nitrogen status is better for the Wet and Intermediate Zone soils than for the Dry Zone soils (see Fig. 8.16 C). Here again climatic influence on organic matter accumulation may be seen. Available phosphorus status is generally not good (see Fig. 8.16, D). In view of the fact that most of these samples were taken at experimental stations, the different grades assigned may show only the difference in intensity of management on the particular plot sampled.

8.9 THAILAND

The samples from Thailand may be considered the most accurate in representing the paddy soils of a particular country. The 80 samples used in this study were selected from 240 samples collected from all over the country, at different times, and correspond to the extent of rice land in different regions. A detailed study of these samples was conducted by Prachak Charoen in his doctoral thesis at Kyoto University (1974). In this book the results of his study were fully utilized, although with the wider aim of placing Thai soils in the paddy soils of tropical Asia. We also used the results of our previous study, *Lowland Rice Soils in Thailand* (Kawaguchi and Kyuma, 1969).

There are five geographical regions: the intermontane basins of the North, the Upper Central Plain, the Bangkok Plain, the Northeast or Khorat Plateau, and the southern peninsula. Climatically Thailand is fairly uniform, the only exception being the southern peninsular region, which is not only wetter but also has a different rainfall pattern. In parent material, the Northeast, quite different from the rest, is

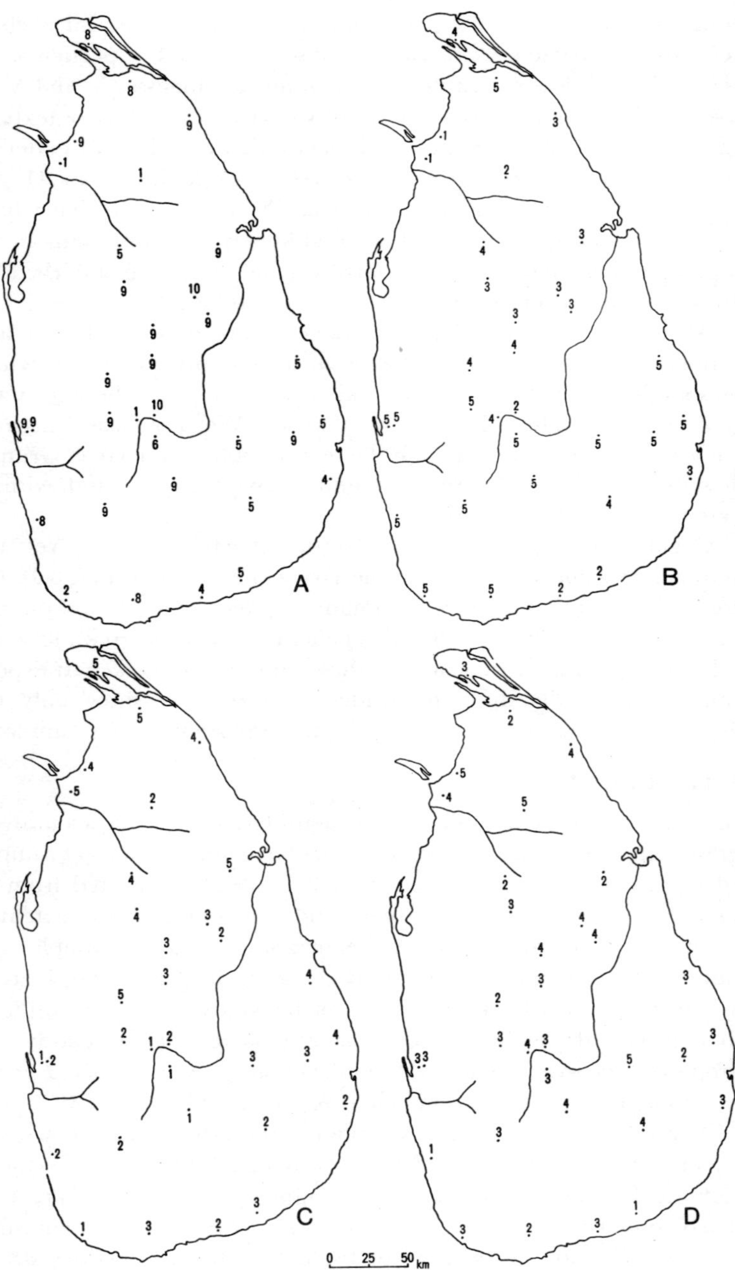

FIG. 8.16   MAP OF SRI LANKA, SHOWING DISTRIBUTION OF SAMPLES
IN TERMS OF: A, MATERIAL CLASS; B, INHERENT POTENTIALITY GRADE;
C, ORGANIC MATTER AND N GRADE; D, AVAILABLE P GRADE.

composed almost exclusively of weathered Mesozoic sandstones, a continuation of the Mesozoic formations in Cambodia.

Physiographically, two major subdivisions are possible in each region, that is, areas of low-lying recent fluvial or deltaic sediments and fan-terrace (or plateau) areas of higher elevation. The relative extent of these two varies from one region to another. The Northeast has a large area of higher elevation, which, together with the parent material, conditions the characteristics of the soil cover.

The sampling sites are plotted on Figure 8.17. The Northeast with the largest rice acreage has the largest number of samples, and next would be the Bangkok Plain. The Upper Central Plain is somewhat important in both acreage and production. The northern intermontane basins and the southern peninsula are of minor importance for rice production, but the former produces many kinds of annual crops as irrigated off-season crops, such as vegetables and pulses, while the latter specializes in perennial commercial crops, such as rubber and fruits.

Dominant soil material classes are VIII in the Northeast, I and III in the Bangkok Plain, and II and III in the other regions (see Fig. 8.18, A). Class I in the Bangkok Plain, and elsewhere, is mostly grumusols and grumusolic alluvial soils. Class III is typically soils on brackish sediments, but also includes some backswamp soils in the Upper Central Plain. Class II is mostly strongly leached fan-terrace soils. Class VIII is sandy, severely depleted material as typically seen in the Northeast. It consists of pink-colored quartz sand and contains very little available nutrient of any kind.

Inherent potentiality reflects the soil material characteristics. Soils with inherent potentiality grade 5 are in the majority in the Northeast (see Fig. 8.18, B). Southern peninsula soils derived from granitic parent materials are often grade 4 or 5. Grumusols and grumusolic alluvial soils have an inherent potentiality of grade 1. Many Bangkok Plain soils are grade 2, and acid sulfate soils are not differentiated from others as far as surface soil is concerned.

Organic matter status is again poor for soils in the Northeast and for fan-terrace soils on the periphery of the Bangkok Plain and the Upper Central Plain, as seen from Figure 8.18, C. The soils in the Bangkok Plain proper and in the southern peninsula have higher organic matter status, reflecting the wetter soil water regime due to physiography and climate respectively.

As shown in Figure 8.18, D, available phosphorus status is generally poor. This is even true of the Bangkok Plain soils. Just one soil, on non-acid coastal marine clay, has an exceptionally high available phos-

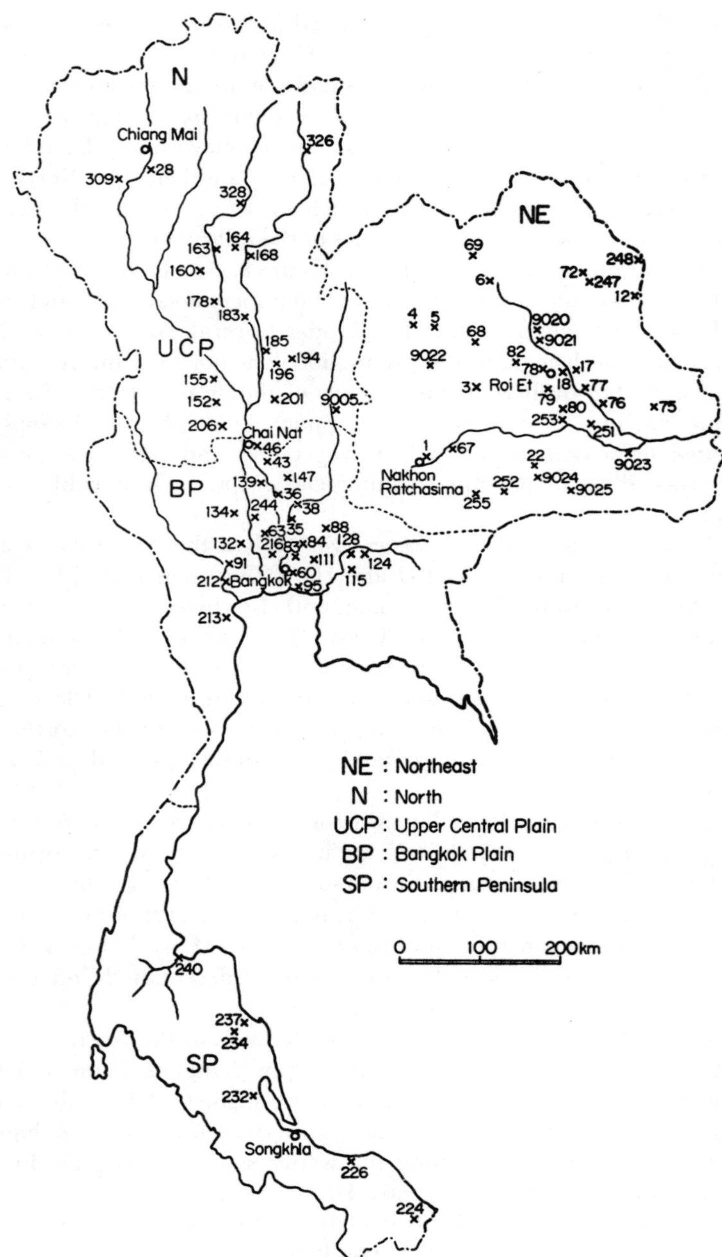

FIG. 8.17   SAMPLING SITES IN THAILAND

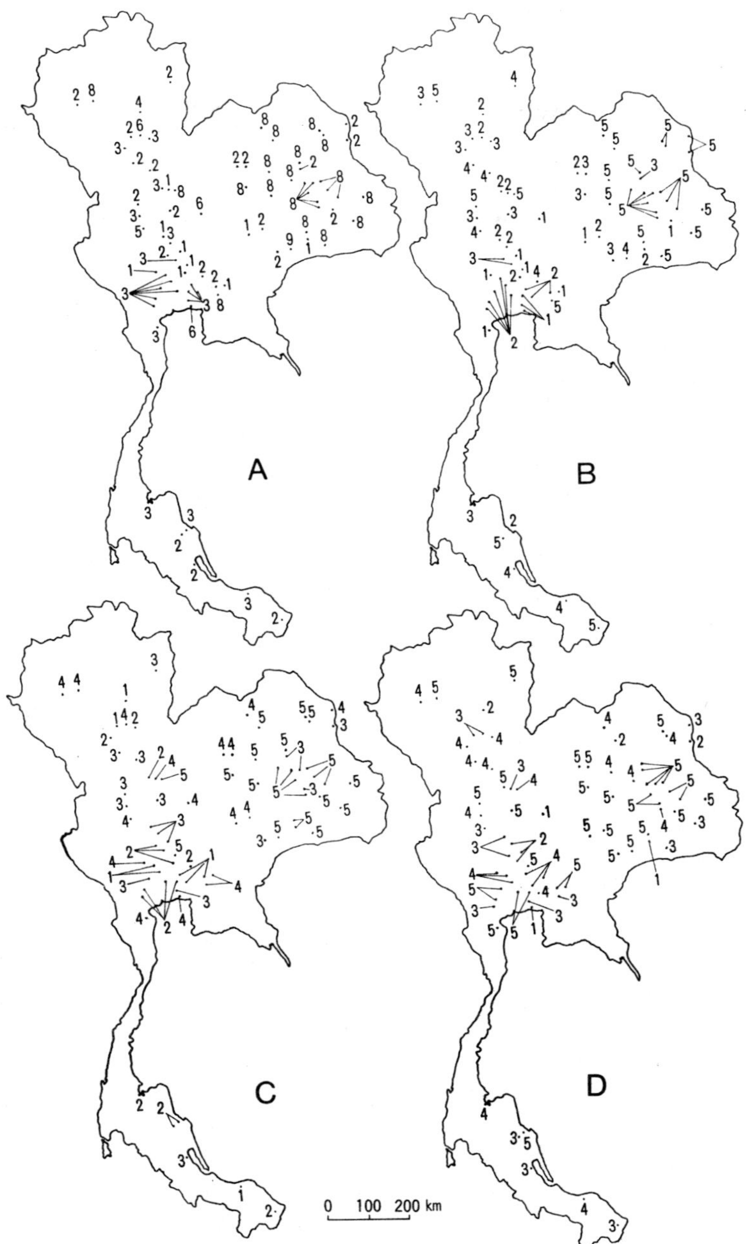

FIG. 8.18 MAP OF THAILAND, SHOWING DISTRIBUTION OF SAMPLES IN TERMS OF: A, MATERIAL CLASS; B, INHERENT POTENTIALITY GRADE; C, ORGANIC MATTER AND N GRADE; D, AVAILABLE P GRADE.

phorus status, grade 1, which is probably due to biological concentration of phosphorus by an unknown mechanism.

8.10 VIETNAM

The Vietnamese part of the Mekong delta has about two million hectares of rice land, which accounts for a very substantial part of total rice acreage in the southern part of Vietnam. The Mekong delta region was surveyed for this study (see Kyuma, 1976), after the preparation of the first draft of this book. Because of the then unfavorable security situation, samples were collected from areas with ready access from main road, thus omitting the Plain of Reeds, the Ha Tien Plain in Kien Giang Province, and the broad depressions in Chuong Thien Province on the right bank of the Bassac River. These unsurveyed areas are known to have large areas of acid sulfate soils, most of which have not been utilized for rice cultivation.

A total of 49 paddy soil samples was collected, using the physiographic map prepared by the Netherlands Delta Development Team (1974) for the Mekong Committee. These samples represent tidal flats along the South China Sea coast, river and estuarine floodplains, river levees, and broad depressions, which presumably are filled-up lagoons.

Parent material classification and fertility evaluation were carried out by extrapolating the newly acquired data into the respective schemes, as developed in chapters 6 and 7.

Figure 8.19 shows the sampling sites. The samples were grouped according to physiographic units as follows:

High tidal flat	V-2, 4, 6, 7, 9, 10, 12, 22, 44, 46, 49 ~ 51
Low tidal flat	V-1, 3, 5, 36, 45, 47, 48
Floodplain	V-13 ~ 16, 19, 20, 27, 30 ~ 32, 37, 39, 41 ~ 43
Levee	V-23, 28, 29, 33 ~ 35,
Broad depression	V-17, 18, 21, 25, 26, 38, 40
Piedmont	V-24

From Figure 8.20, A, extreme uniformity of soil materials is at once clear. The two class VIII samples in the northwestern part of the delta are V-24 on a piedmont of granite hills and V-26 on outwashes from nearby granite hills. The class II sample, V-22, is underlain by a layer containing large iron nodules that cement sandy subsoil materials and thus is better drained. The widely occurring class III materials are fine textured and poor in bases. The mean pH of 4.5 and the fact that 86 percent of the samples have heavy clay texture explain the predominance of class III in the Mekong delta. Because our samples include only a few soils of acid sulfate nature, the low pH is not necessarily ascribable to acid sulfate soils. Even soils on the Mekong and Bassac levees show quite low pH around 5.

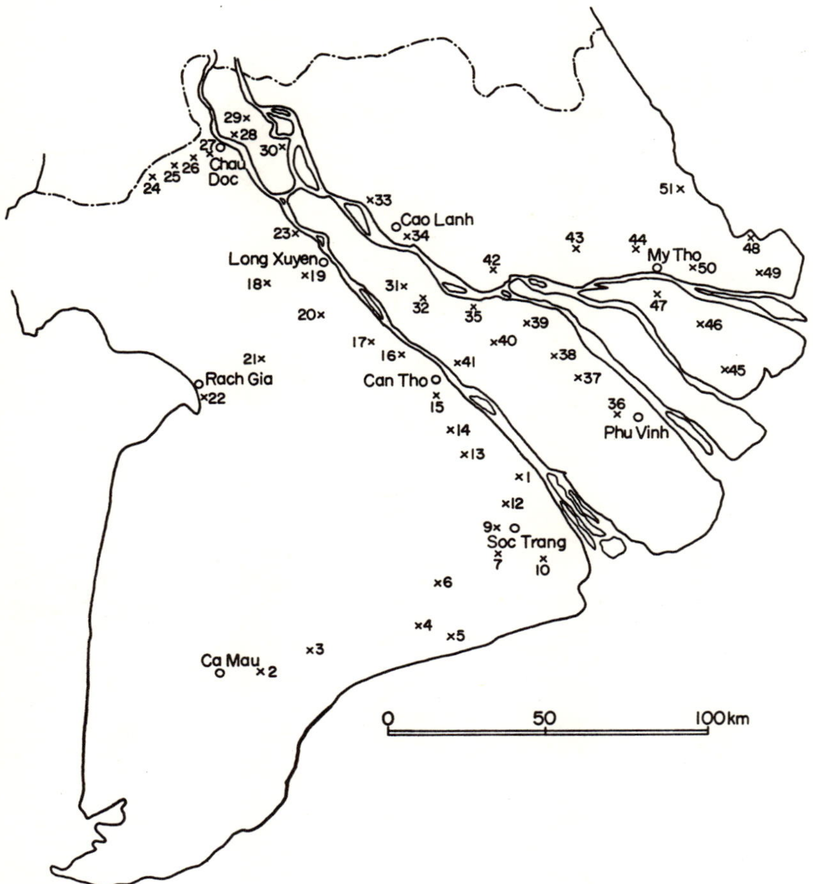

FIG. 8.19   SAMPLING SITES IN THE MEKONG DELTA, VIETNAM

Figure 8.20, B shows that inherent potentiality grades for the sample soils are also quite uniform. The soil on the granite piedmont has a very low potentiality, but others are either intermediate or high in inherent potentiality, grade 2 or 3. Figure 8.20, C shows organic matter status, which is generally high, the one exception being the piedmont soil. Soils on high tidal flats and river levees tend to be a little poorer in organic matter status. Available phosphorus status is not good for the Mekong delta soils, as is apparent from Figure 8.20, D. Many soils are grade 4 or 5. In this case, the levee soils appear to be little better than others.

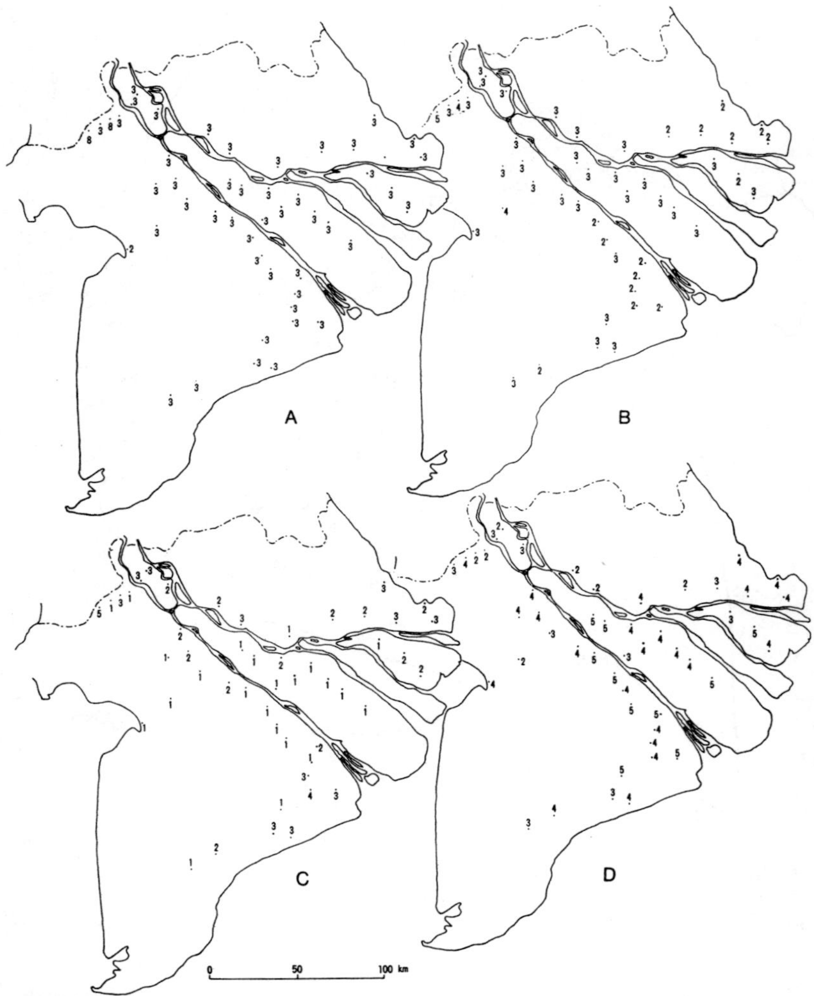

FIG. 8.20   MAP OF THE MEKONG DELTA, VIETNAM, SHOWING
DISTRIBUTION OF SAMPLES IN TERMS OF: A, MATERIAL CLASS; B, INHERENT
POTENTIALITY GRADE; C, ORGANIC MATTER AND N GRADE; D,
AVAILABLE P GRADE.

9

Specific Problems Related to Paddy Soils in Tropical Asia

So far we have discussed the problems of paddy soils in tropical Asia using the results of our own studies of samples taken in different countries in the region. There remain, however, such topics related to paddy soils, especially to their fertility, as silting, acid sulfate soils as a potential medium of rice cultivation, nutrient deficiencies and toxicities in relation to plant performance, etc. In this chapter we deal with these topics in the light of the studies recently carried out in tropical Asia and in Japan.

9.1 SILTING

In the long run there is little doubt about the positive effect of silting in maintaining soil fertility. This point was made in chapter 2 in relation to the advantages of rice cultivation over upland crop cultivation in tropical Asia. If water conditions are adequate, the superiority for rice cultivation of alluvial soil, which is, in effect, accumulated silt, would last long after annual addition of silt ceases. Here we will discuss not this long-term effect but the short-term effect of silting on the maintenance of desirable conditions for rice cultivation.

The famous words of Herodotus that Egypt is a gift of the Nile illustrate the widespread belief that river silt brings to the soil something which helps farmers cultivate the same piece of land year after year. This something is, first and foremost, plant nutrients. As stated in chapter 3, river water sometimes brings a significant amount of mineral nutrients, such as calcium, magnesium, potassium, and silicon, to a soil, but only a negligible amount of phosphorus and nitrogen relative to crop requirements. And in paddy soils, nitrogen can be supplied by microbial fixation, as mentioned in chapter 3. Therefore, if there is anything positive expected of silting, it must be a supply of phosphorus, and, though much less important, potassium and other nutrient cations, and silicon.

To examine this point, first, we have to estimate the amount of silt deposited on paddy fields by the annual flood. The sediment load in river water at the high water season often exceeds 300 ppm, but much of this is carried down to the sea. No close estimate is available for the amount of silt added yearly to paddy fields. But the following observation by Brammer (1964) in East Pakistan (now Bangladesh) appears relevant; "Flying over the Ganges and Brahmaputra floodplains at the height of the 1962 floods, the writer was struck by the fact that only near the major river channels did the flood water appear silty. Away from the rivers, over the greater part of the floodplain, the water appeared clear."

The Netherlands Delta Development Team (1973) estimated silting in the Mekong delta of Vietnam. It noted that, even if all the silt load of the river were deposited in the delta, the annual addition would range from 0.2 to 0.7 mm, with flood depths of 1 to 3 m. Another of the team's estimates was based on the geomorphological consideration that the 1 to 2 m of recent alluvial deposits, overlying potentially very acid sediments in the estuarine floodplain, had accumulated over the past 3500 years. This gave an average of 0.3 to 0.6 mm accretion per year. Even further upstream, in the river floodplain in Cambodia, the annual accretion was estimated at about 1 mm per year.

Uehara *et al.* (1974) recently conducted a study of the composition of Mekong river silt to evaluate its possible role in delta soil; and, on the assumption of 1 mm of silt deposition per year, concluded that nutrient elements supplied would be no more than 1 kg of P, 3.2 kg of K, 4 kg of Mg, and 50 kg of Ca per hectare and that these amounts are negligibly small in comparison with the element contents of the surface 10 cm of the delta soils.

The addition of nutrient elements, particularly of P, by silting is not significant and, thus, the general belief in the short-term effect of silting is discredited. At the same time, however, the example just given allowed us to confirm the long-term effect of silting. One mm per year of silt builds up in 100 years to 10 cm of surface soil, which was the plow-layer depth most commonly observed in our surveys of tropical Asian paddy soils. Even where the accumulation of 10 cm takes ten times as long, the effect of silting on soil rejuvenation still should be regarded as positive.

There are certain areas where this rejuvenation effect is seen on a shorter time scale. In Vietnam one sample (V-33) was taken about 1 km inland from the Mekong River along a canal dug some sixty years earlier. The profile showed clearly two different sediments, the upper 30 cm consisting of reddish colored (5YR–7.5YR) silty material, similar to Mekong levee material, which is underlain by grayish colored (5Y–7.5Y) clayey material, the top of which is stained by organic matter. It is obvious that the lower material was formerly on the surface in a backswamp. Since the construction of the canal, the upper sediment has been deposited by the Mekong floods. As a result the surface material has a relatively high amount of exchangeable Ca and available phosphorus and contains no detectable sulfur, whereas the subsoil material is more depleted and contains about 100 ppm of total sulfur. This example can be compared with "colmatage" irrigation practices in Cambodia and illustrates short-term soil rejuvenation by silting. Thus, in the upstream areas of the major rivers of tropical Asia such effects can be expected along river channels and newly dug canals. But this observation cannot be generalized to confirm the positive short-term effect of silting in the entire area of the delta.

9.2 ACID SULFATE SOILS

Actual and potential acid sulfate soils occur extensively in tropical Asia, especially in the deltas and coastal lowlands of Southeast Asia, and these areas are considered to be important potential agricultural lands.

In the Mekong delta of Vietnam more than one million hectares of acid sulfate soils are known to occur, notably in the Plain of Reeds, the Ha Tien Plain of Kien Giang Province, and in Chuong Thien Province, in a broad depression southwest of the Bassac River. Most of these areas have not been utilized for agriculture. Cambodia also has a small area of acid sulfate soils at the margins of the Plain of Reeds. These soils were designated as Alumisol in the soil map compiled by Crocker (1962).

According to van der Kevie (1972), Thailand has about 800,000 hectares of acid sulfate soils in the Bangkok Plain alone with more in the Southeast and in the peninsula. But since van der Kevie defined acid sulfate soils as those having a cat-clay horizon somewhere in their profile, not all these areas are strongly acid and the greater part has surface sediments with tolerable acidity which can be and are being utilized for rice growing.

In West Malaysia about 110,000 hectares of acid sulfate soils occur, mainly on the west coast, of which 25,000 hectares are under rice and approximately the same number under rubber. Other crops—such as oil palms, coconuts, and vegetables—are grown on these soils, but there are about 45,000 hectares of unused land (Kanapathy, 1973). In East Malaysia, especially in Sarawak, there is an extensive coastal swamp area, which has either peat or potential acid sulfate soil quite near the surface. No exact estimate is available, but there are at least a few hundred thousand hectares of acid sulfate soils in Sarawak and parts of Sabah.

For Indonesia, Driessen and Soepraptohardjo (1974) recently reported that about two million hectares of (partly potential) acid sulfate soils are present along the coasts of Sumatra and Kalimantan. West Irian also has quite a large area of sulfide-containing sediments along the coast, but they are potential acid sulfate soils or mud clays.

Various scholars indicate the occurrence of acid sulfate soils in North Vietnam (Fridland, 1964), Burma (Kyuma, unpublished), the Philippines (Tanaka and Yoshida, 1970), and Bangladesh (Islam, 1955; Brammer, 1970), but their extent in any one country is not very large. Sri Lanka and India have practically no acid sulfate soil problems.

As the morphological, biological, chemical, and physical aspects of acid sulfate soils have been discussed by many researchers from different parts of the world (see the papers edited by Dost, 1973; Bloomfield and Coulter, 1973; van Breemen, 1976), we will not repeat these here. However, we will consider the management problems of acid sulfate soils for rice cultivation.

First, we refer to the results of our study on acid sulfate soil samples (Ca-13, M-9, M-27, T-63, T-84, T-88, V-25, and V-47). In terms of material and fertility characteristics, these soils are not conspicuously different from the other deltaic soils. They mostly fall into soil material classes II and III, which are clayey, more or less siliceous, and base poor materials commonly found in Southeast Asian deltas. In terms of fertility components these acid sulfate soils are slightly inferior in inherent potentiality (grade 2–3), usually richer in organic matter and nitrogen status (grade 1–2), and slightly poorer in available phosphorus status (grade 3–5), than the non-acid sulfate soils occurring in the same area. This lack of discrepancy may be because we dealt with the surface soil, which is not necessarily of an acid sulfate nature. Another possibility is that the method, that is, the computation of the factor score on an additive model, disabled one or two extreme characters to conspicuously affect the final score. But, we feel the result is not necessarily unreasonable. Acid sulfate soils which have been under cultivation for a considerable time, such as our samples, hold their own even in terms of paddy yield. They seem to have been ameliorated in the course of cultivation.

Therefore, we confine our present discussion to the problems encountered when newly reclaimed acid sulfate soils are to be used for rice cultivation.

Because acid sulfate soils occur in swampy areas, when they are to be utilized for agriculture, rice is often the first crop to be considered. In addition to its tolerance of swampy conditions, rice can also tolerate the fairly high salinity to which it is apt to be exposed in acid sulfate soils. Electric conductivity up to 4 mmho/cm is said not to affect rice growth adversely even at the transplanting stage, when rice is most susceptible to salt injury.

Rice is also one of the crops that can tolerate relatively high acidity. Moreover, the pH of even acid sulfate soils tends to rise in submerged conditions, though not as much as in ordinary soils. Extreme acidity, say pH below 4, activates aluminum. In some studies conducted in Japan (Takijima, 1963; Takahashi, 1961) it was found that rice seedlings start to show growth inhibition when Al^{3+} ion concentration exceeds 35–40 ppm, which concentration may be encountered in acid sulfate soils with partial oxidation and restricted drainage.

Moreover, ferrous sulfate, produced in quantity in the oxidation process of pyrites,* is said to adversely affect the growth of rice (Kobayashi, 1939); rice shows symptoms of ferrous sulfate injury, that is, blackish brown coloring of the tip of the leaves, and finally dies. The injury is enhanced when H_2SO_4 is also present. Takijima (1963) experimentally showed that root elongation of rice seedlings was inhibited in half the experimental plants, when the concentration of ferrous sulfate exceeded 150 ppm.

Therefore, although the low pH itself can be tolerated, the concurrent reactions producing high concentrations of ferrous sulfate and active aluminum could cause the physiological disorders of rice plant in acid sulfate soils. Another side effect of low pH is the reduction in available phosphorus, which is induced by the high activity of aluminum. Thus, phosphorus deficiency is often reported in acid sulfate soils.

The following three measures are conceivable means of alleviating the harmful concentrations of toxic substances:

(1) not to allow oxidation of pyrites contained in the potentially acid sulfate sediments
(2) to leach the harmful substances out of the rooting zone after allowing oxidation to occur
(3) to inactivate aluminum by raising the pH of the medium by liming, which also reduces ferrous iron concentration in the soil solution.

---

*According to Temple and Koehler (1954) the oxidation process of pyrites can be summarized as follows:

$$FeS_2 + H_2O + 7\,O = FeSO_4 + H_2SO_4 \text{ (chemical process)}$$
$$2\,FeSO_4 + O + H_2SO_4 = Fe_2(SO_4)_3 + H_2O \text{ (by } Thiobacillus\ ferrooxidans\text{)}$$
$$Fe_2(SO_4)_3 + FeS_2 = 3\,FeSO_4 + 2\,S \text{ (chemical process)}$$
$$2\,S + 6\,Fe_2(SO_4)_3 + 8\,H_2O = 12\,FeSO_4 + 8\,H_2SO_4 \text{ (chemical process)}$$
$$S + 3\,O + H_2O = H_2SO_4 \text{ (by } Thiobacillus\ thiooxidans\text{)}$$

All three measures have been recommended by researchers who have dealt with acid sulfate soil problems in relation to rice cultivation.

The first measure is not a positive solution of the acid sulfate soil problem. It suppresses the natural ripening process of the so-called mud clay (potentially acid sulfate sediments). First of all, because of strongly reductive conditions and very unfavorable working conditions, rice yield remains very low, and yet no positive measures for yield improvement can be taken. In addition, this apparently easy solution is not actually easy, because of unpredictable climatic fluctuation. When unusually long dry spells occur, the soil surface is inevitably oxidized and strongly acidified, thus leading to total failure of the rice crop. Tanaka and Yoshida (1970) reported such an incidence in Camarines Sur, the Philippines.

The second measure can be taken only where good drainage facilities are provided. In one lysimeter experiment conducted by Murakami (1965) in Japan, in a heavy clay soil (50 percent clay and 42 percent silt) plot that was provided with tile drainage at a depth of 80 cm, the third crop (one crop a year) gave over 4 tons/ha of paddy, whereas a plot without drainage gave almost no yield through the first four crops and only after that was the yield remarkably increased. In the first three years of percolation treatment, oxidizable sulfur content decreased from the initial 19.2 mg to 1.2 mg per 100 g soil. This area has circa 800 mm of precipitation during the fallow months, and this must have contributed greatly to the removal of toxic substances through natural percolation.

We must consider here the conditions under which acid sulfate soils occur in Southeast Asia. Most of the areas are low-lying, and natural drainage is severely restricted. During the rainy season, the entire land is submerged, but during the dry season many of the areas become dry to a depth of about thirty centimeters. Surface water during the rainy season is mostly fresh and does wash the surface few centimeters of soil, but in the dry season the upward capillary movement of water cancels out this effect, bringing the toxic substances back to the surface.

The minimum requirement for reclamation of acid sulfate soil areas is to dig open ditches for drainage, except when the first measure of not allowing oxidation is being taken. Often tidal gates have to be installed in coastal areas to prevent saltwater intrusion during the dry season. This investment certainly reinforces the washing effect during the rainy season, but it is not effective at all during the dry season unless there is an ample supply of freshwater, which is difficult to attain in a monsoonal climate.

One passive countermeasure to the upward movement of toxic

substances during the dry season is mulching the soil surface. If it reduces the capillary rise of water during the dry season, the effect of washing by the early rains and during the rainy season would be accumulated and thus the surface soil condition could be improved rapidly. Since, as Murakami (1965) and others pointed out, rice requires only a shallow rooting zone, say 10 to 15 cm, to produce a moderate yield, even the minor work of mulching should accelerate the establishment of paddy fields. With open ditches, tilling of the surface soil toward the end of the dry season would also be effective. But for this, large tractors and firm ground are necessary, and these requirements are not always satisfied in acid sulfate soil areas.

In areas with a permanently humid climatic condition, as typically seen in Sarawak, East Malaysia, leaching first with sea or brackish water followed by that with plenty of rainwater could be an effective means of reclaiming acid sulfate soil. Ridging, or, to the same effect, a dense network of shallow ditches, should also prove effective in such areas for upland crop cultivation, provided major drainage works are furnished.

The third measure, liming, is effective in raising soil pH and thus inactivating toxic aluminum. But if this measure alone is taken, the amount of lime required would easily amount to ten tons per hectare or more, which is hardly economical. Moreover, the difficulty of transporting the lime to the initially swampy land must not be forgotten.

Another point to be considered is that oxidation of pyrites, which are the dominant form of oxidizable sulfur in potentially acid sulfate soils, is retarded in a neutral reaction range (see Murakami, 1965). Therefore, if lime is applied at the beginning of reclamation, the time required to leach out the toxic products of oxidation is prolonged.

Thus, liming should be carried out after the effect of oxidation and leaching becomes apparent. In Thailand, liming is recommended in amounts just enough to inactivate aluminum (Komes, 1973). There, further liming to produce higher pH is thought to have the adverse effect of causing sulfate reduction in the rooting zone. Murakami's experiment (1965), however, indicates that in acid sulfate soils harmful effects of hydrogen sulfide because of sulfate reduction are rather rare. He argues that in these soils ferric oxides are liberated and precipitated as the final product of oxidation of pyrites, and, therefore, the redox potential is kept relatively high, high enough to suppress sulfate reduction.

When the latter measures of reclamation, that is, oxidation and leaching with or without liming are taken, the loss of soil mineral nutrients, and, accordingly, the lowering of soil fertility due to acid leaching are worrisome. Murakami (1965), however, concluded from his experiments that such effects are not serious. His results revealed that : (1) all the exchangeable cations decreased remarkably, but Ca, Mg, and K in the soil solution are kept at a level high enough to supply rice plants with these elements; (2) cation exchange capacity did not change appreciably; (3) phosphorus absorption capacity did not increase; and (4) the C/N ratio of soil organic matter was lowered from about 20 to 10, at the end of three cropping seasons. Further studies are necessary before a definite conclusion can be drawn.

9.3 NUTRIENT DEFICIENCIES AND TOXICITIES

This subject was dealt with by Tanaka and Yoshida (1970) in one of the publications of International Rice Research Institute (IRRI). They surveyed the literature and made field studies in Asian countries and counted the following as causes of nutritional disorders in rice plants: deficiencies of phosphorus, potassium, zinc, iron, manganese, and silica, and toxicities of iron, aluminum, manganese, and sulfide. In addition, the toxic effects of organic acids, carbon dioxide, and salinity were noted. A summary is given in Figure 9.1 and Table 9.1. We will first take up the problems most frequently encountered in tropical Asia, that is, phosphorus deficiency and iron toxicity.

Phosphorus Deficiency

The relative advantage of paddy fields over upland conditions in availability of phosphorus was discussed in chapter 3. Even with this advantage and even at the present low yield level of the region, many paddy soils are still deficient in phosphorus. As shown in chapter 5, nearly a quarter of our samples contain less than 40 mg of total P_2O_5 per 100 g of soil, which is thought to be critically low for normal rice growth. In terms of available phosphorus measured by the Bray-Kurtz No. 2 method, more than 50 percent of these soils contain less than 15 ppm of P_2O_5. These figures endorse Tanaka and Yoshida's findings of widespread phosphorus deficiency.

The phosphorus absorption coefficient, which was measured by a method (see Appendix 2) routinely used in Japan, is usually moderate to low (less than 1500 mg P_2O_5/100 g soil) and does not seem to be a serious problem. Only the soils containing free lime and ando soils show high phosphorus absorption coefficients (1500–2500 mg P_2O_5/100 g soil). Even latosols rich in active iron oxides do not show particularly high values. Moreover, if iron is absorbent, phosphorus availability would not be seriously lowered in paddy field conditions, as discussed in chapter 3. Therefore, it seems that the main cause of the low available phosphorus status is the low absolute phosphorus content.

The fertility index, available phosphorus status (AP), derived from chapter 7, gives summarized information on phosphorus availability of paddy soils. The soils of AP grade 4 and 5 are liable to be deficient in phosphorus, and these soils are widespread, especially in Thailand and Cambodia. In the east coast of West Malaysia, West Java, and the Mekong delta of Vietnam, there are also many soils with low AP grades. Even soils with high AP grades require application of phosphatic fertilizers if a higher yield is desired.

Iron Toxicity

Iron toxicity seems to occur in two groups of soils, one being acid sulfate soils and the other strongly depleted soils of a latosolic nature. As already stated, the former is widely distributed in Southeast Asia, but the occurrence of the latter is rather limited as far as paddy soils are

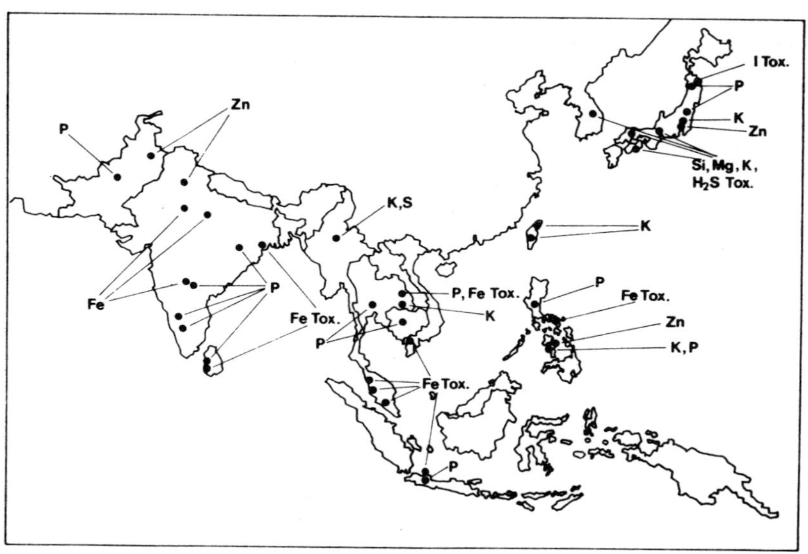

LEGEND
P = Phosphorus deficiency
K = Potassium deficiency
Fe = Iron deficiency
Si = Silicon deficiency
Mg = Magnesium deficiency
S = Sulfur deficiency
Fe Tox. = Iron toxicity
I Tox. = Iodine toxicity
H₂S Tox. = Hydrogen sulfide toxicity

Fig. 9.1   NUTRITIONAL DISORDERS OF RICE IN ASIA (TANAKA AND YOSHIDA, 1970)

TABLE 9.1 CLASSIFICATION OF NUTRITIONAL DISORDERS IN ASIA

SOIL	SOIL CONDITION	DISORDER	LOCAL NAME
Very low pH	(Acid sulfate soil)	Iron toxicity	Bronzing
Low pH — High in active iron	Low in organic matter	Phosphorus deficiency	
	High in organic matter	Phosphorus deficiency combined with iron toxicity	
	High in iodine	Iodine toxicity combined .... with phosphorus deficiency	Akagare Type III
	High in manganese	Manganese toxicity	
Low pH — Low in active iron and exchangeable cations	Low in potassium	Iron toxicity interacted with ...... potassium deficiency	Bronzing / Akagare Type I
	Low in bases and silica, with sulfate application	Imbalance of nutrients associated with hydrogen sulfide toxicity	Akiochi
High pH	High in calcium	Phosphorus deficiency / Iron deficiency / Zinc deficiency	Khaira / Hadda / Taya-Taya / Akagare Type II
	High in calcium and low in potassium	Potassium deficiency associated with high calcium	
	High in sodium	Salinity problem / Iron deficiency / Boron toxicity*	

*Probably rare.

From Tanaka and Yoshida (1970)

concerned. Bronzing, which is thought to be a symptom of iron toxicity, occurs in Sri Lanka in a small valley filled with sandy sediments from surrounding low hills covered with latosolic soils. One of our samples Sr-27 was taken in such an area. Typical iron toxicity was observed by Tanaka and Yoshida (1970) "on sandy lateritic soils at Bhubaneswar, Orissa State" in India. This is near our sample I-27 and in an area with many abandoned laterite quarries.

The description of iron toxicity by Tanaka and Yoshida is somewhat controversial. The symptoms of iron toxicity are not directly connected with a high iron content in the medium nor in the plant. They seem to regard it as a result of interaction between a high iron content in the soil and deficiency of one or more of such essential nutrients as potassium and phosphorus. According to Tadano (1975), rice plants have a built-in protective mechanism which enables them to exclude an excessive amount of ferrous ions, if they are nutritionally healthy, especially with respect to potassium, magnesium, calcium, and phosphorus.

In the case of latosolic soils, such as those observed in Sri Lanka and India, the status of all nutrients is normally low. In addition, active iron content is also often very low. Under such a condition hydrogen sulfide evolved in the reduction process would cause inhibition of the respiratory function of the plants. Since the protective mechanism of the rice plant just mentioned is associated with respiration, sulfide injury can fatally affect the mechanism and induce excessive iron uptake, which causes the iron toxicity, even with a relatively low concentration of ferrous iron in the soil. In fact, Yamada (1959) and Inada (1966) suggested, in their studies on bronzing in Sri Lanka, that root damage by hydrogen sulfide is a possible primary cause of iron toxicity.

In acid sulfate soils the situation may be different. First of all, an extremely low pH due to free sulfuric acid produced by pyrite oxidation would weaken the protective mechanism of rice roots against the uptake of an excessive amount of iron. Furthermore, a poor nutritional status of the plant aggravates the toxicity. In this connection a generally low phosphorus level in acid sulfate soils is most worrisome. A high salinity due to NaCl or $MgCl_2$ is known to decrease the oxidizing power of rice roots and thus enhance iron toxicity (Tadano, 1975). In many young acid sulfate soils iron toxicity partly due to this cause could be widespread.

Although much has been clarified about iron toxicity, further studies are still needed to be able to effectively counteract the disorder. The real cause of what was called ferrous sulfate injury by Kobayashi (1939), Takijima (1963), and Murakami (1965) for the rice grown in acid sulfate soils (see the section on Acid Sulfate Soils) must also be studied in relation to iron toxicity.

Other Nutrient Deficiencies

Generally speaking, nutrient deficiencies are mostly due to the following: low absolute amount, or low availability or solubility, of the nutrients concerned. Antagonism between two or more nutrients is another, less frequent, cause.

As discussed in chapter 3, contents of various nutrients are mutually highly correlated and, thus, a soil that is very low in one nutrient tends to be low in others also. Therefore, deficiencies from the first cause, that is, low absolute amount, often occur simultaneously for many nutrient elements. In an acid sandy soil with low CEC and low exchangeable cations—potash, calcium, magnesium, manganese, silica, and other micronutrients may be simultaneously deficient, but, of course, by one of these at the lowest availability level the yield is limited. Of the soil samples we studied, those of inherent potentiality grade 5 are most likely to exhibit such multiple nutrient deficiencies. Soils in Northeast Thailand and on terraces in Cambodia are notable examples. They are deficient in almost everything. But this type of deficiency is less serious in irrigated fields than in rain-fed fields, because the supply of potassium, calcium, magnesium, and silica from irrigation water is considerably large.

Sulfur deficiency, however, is independent. It usually occurs inland, away from the sea. Sulfur deficiency is suspected in what is called yellow leaf disease in Upper Burma (Tanaka and Yoshida, 1970), and in a disorder found in grumusol and planosol areas of Java (Ismunadji *et al.*, 1975).

Silica is not usually counted among the essential elements for upland crops, but for rice a certain high level seems to assure healthy growth and a high yield (Takahashi, 1961). As stated in chapter 5, threshold values of available silica both for rice plants and soil are known for Japanese conditions. But these same values may not be applicable to tropical Asian conditions. There are a few studies on the silica content of tropical rice plants, in which 8–10 percent of dry matter is regarded as the approximate threshold content of silica (Yoshida *et al.*, 1969; Takijima *et al.*, 1970). There are no reliable data for setting a threshold value for soil silica in tropical Asia. As the rate of silica release from soil minerals would be much faster, the static threshold value would be lower than that for Japan. Even so, there could be many silica deficient soils, as revealed in our previous studies (Kawaguchi and Kyuma, 1969a, 1969b), and these soils are mostly among those with an inherent potentiality of grade 5. The peat soils being rapidly reclaimed for rice cultivation, especially in Indonesia, are another category of soils that would respond to available silica application, although no studies proving this have been carried out.

The second cause of nutrient deficiency, that is, low availability or solubility, is often seen in soils with a high pH. The solubility of iron and manganese is markedly reduced in the neutral to alkaline reaction range, and the availability of zinc is conditioned by a high concentration of bicarbonate anions in the soil solution in submerged high pH soils. The incidences of zinc and iron deficiency reported by Tanaka and Yoshida (1970) are almost exclusively found in soils containing free lime under drier climatic conditions. High pH soils among our samples with an inherent potentiality of grade 1 are apt to show such deficiencies.

In peat soils, copper deficiency is sometimes suspected to be responsible for a poor performance of rice, though we have not come across any verification in the literature. Polak (1951) reported an increased yield of maize as a result of application of copper sulfate in a potted experiment with a peat soil. This is an example of low solubility of copper induced by chelation.

An example of the third cause of nutrient deficiency, that is, antagonism, is seen in potash deficiency in soils containing a high amount of calcium and/or magnesium. But this seems to be not a very important cause of potash deficiency in tropical Asia. Soils with an inherent potentiality of grade 1 are more likely to exhibit this trouble.

9.4 *AKIOCHI* OR DISORDERS RELATED TO STRONG REDUCTION

Akiochi is a Japanese word meaning autumnal decline in plant activity. In Japan with heavy doses of fertilizers, rice plants show vigorous growth until midsummer, but toward the end of summer this growth declines and results in a very unsatisfactory yield. The type of soil that shows this *akiochi* phenomenon is a whitish colored sandy soil, the profile characteristics of which are similar to those of M-36 as described in chapter 3. The most prominent feature is the presence of an A2-like bleached horizon, though not all the degraded *akiochi* type soils have this feature.

The cause of *akiochi* is linked with a low, free iron oxides content, but this does not mean physiological iron deficiency. High summer temperatures accelerate the process of decomposition of organic matter, causing a strong reduction in the plow layer. Of the anaerobic decomposition products, butyric acid and hydrogen sulfide are particularly toxic to rice plants. If the content of free iron oxide is kept sufficiently high, the redox potential is buffered at a relatively high level and anaerobic fermentation, which produces butyric acid, and sulfate reduction would be suppressed. Even if sulfide is locally produced, it would be captured by ferrous iron to precipitate as ferrous sulfide. Therefore, the remedy for *akiochi* soil is, first of all, to incorporate iron-rich red-colored soil materials into the plow layer (soil dressing), and by so doing, other nutrient elements, such as silica and magnesium, are added to the soil, which is likely to be deficient in these elements also. The threshold content for Japanese paddy soils is set at 1 percent Fe_2O_3 as determined with the Mg-ribbon reduction method (Kawaguchi and Matsuo, 1955), below which the *akiochi* phenomenon is likely to occur.

Our previous studies (Kawaguchi and Kyuma, 1969a, 1969b) revealed that many soils from Northeast Thailand and West Malaysia have free iron content below this level. Though the threshold content of free iron oxide has yet to be checked for applicability to tropical conditions, it is recommended that such iron-poor soils avoid the use of fertilizers containing sulfur and that the soil conditions are not allowed to become strongly reductive.

9.5 SALINITY

Injury of rice caused by salinity is not common in paddy areas. Only in Pakistan and northwestern India, with an arid to semiarid climate, could salinization be a serious problem. If, however, there is an ample supply of high-quality irrigation water, rice should be the most profitable crop even in such soil and climate conditions, because it can stand to a fairly high salinity and/or alkalinity and produce a high yield with abundant sunshine, and at the same time irrigation exerts reclamation effects on the saline and sodic soils.

In the rest of South and Southeast Asia, under a more humid climate, occurrence of salinity problems is confined to coastal areas where saltwater intrudes during the low water season. In the former tidal flats of the Mekong delta of Vietnam, for example, this type of salinity retards the start of the cropping season; farmers have to wait until rains flush out the salt that was accumulated in the plow layer during the dry season.

One special type of salinity problem is seen in Northeast Thailand. In certain areas salt-bearing beds intercalated in the Mesozoic sandstone formation come close to the surface and salt enters the solum by capillary movement (Sinanuwong and Takaya, 1974a,b). The high NaCl content of river water in Northeast Thailand (see Table 3.3) is also due to these salt-bearing beds. In fact, the people of these areas produce and sell crude table salt.

10

Conclusion

The discussions in the preceding chapters have been centered on characterization of the paddy soils of tropical Asia in terms of their material nature and fertility. As the main points of these discussions have already been given, we will not repeat them here. Instead, before closing the discussions, we wish to refer back to the thesis presented in the introduction of this book that soils have to be studied as the basis of intensification and diversification of agriculture. Obviously, we cannot give any definite prescriptions for solving the many soil problems encountered in the course of agricultural development, but we will try to locate the areas where future research is necessary for more intensive use of paddy soils in tropical Asia.

10.1 THE SOIL IN INTENSIFICATION OF RICE CULTIVATION

As discussed in chapter 2, rice cultivation has been developed as an adaptation to the natural environment of the region. It has great advantages over upland crop cultivation in that the use of vast areas of naturally inundated or flooded land is made possible, maintenance of soil fertility is easy, and erosion is minimized. Even with the advanced technology of today these three advantages remain. Therefore, intensification of agriculture in tropical Asia means intensification of rice cultivation.

(a) Water control

In the greater part of tropical Asia, intensification of rice cultivation has not taken place primarily because of difficulties of water control. Therefore, the most crucial factor in achieving intensification is to provide the land with better water control. Without water control, high-yielding varieties that are short-stemmed and lodging-resistant cannot be introduced in such areas of deep flooding as deltas and river flood plains, while in areas without assured water supply, fan-terraces and plateaus, fertilizers cannot be applied because of the economic risk.

In recent years some farmers in the Mekong and Chao Phraya deltas have been cultivating an improved-yield short-term variety of rice during the dry season, by pumping water from creeks and canals. In this way the double-cropping of rice, a long-term native variety and a short-term improved, may become possible in places where floodwater recedes rapidly. This type of intensification is already taking place without major water-control measures. However, this practice of taking irrigation water during the dry season can affect the coastal areas of deltas adversely, by increasing saltwater intrusion. Thus, monitoring of the salinity of river water and the study of soil salt regimes become necessary. When efficient water-control measures, with big dams, are installed, the dry season flow can be increased and stabilized, and this, together with a lowered flood level, would enable an expansion of the double-cropping area without causing problems downstream.

One soil problem, in relation to flood control in the deltas, is the decrease of silting. We discussed silting in chapter 9 and concluded that the short-term effect of silting on soil fertility can be ignored. Of course, in the long run this would result in an appreciable decline of soil fertility, and some countermeasures would have to be taken.

A stabilized water supply is necessary for fan-terrace and plateau areas before any intensification can take place. When irrigation is introduced to areas with drier climates, care must be taken to alleviate salinization. But in the greater part of tropical Asia maintenance of soil fertility is the more important concern when rice cultivation is intensified by irrigation.

(b) Maintenance of fertility

With better water control and higher-yielding improved varieties, soil fertility must be reinforced to give the maximum return. The nutrient balance, established in traditional rice cultivation by natural supplies of nutrient through irrigation water and microbial nitrogen fixation, can be maintained only at yields below 1.5 tons of paddy per hectare. In Japan the nutrient requirements for 1 ton of brown rice, or roughly 1.3 tons of paddy, are said to be 20 kg of N, 10 kg of P_2O_5, and 20 kg of K_2O (Yamane, 1975). If 3 to 4 tons of paddy per hectare is set as a target yield, at least 50 kg of N, 25 kg of P_2O_5, and 50 kg of K_2O are needed for one crop and the use of fertilizers becomes compulsory.

In relation to nitrogen application, disposal of rice straw warrants more careful study. Traditionally, most rice straw is left in the field, some of which is eaten by buffaloes and cattle but the greater part of which is burnt. There may be some justification for this practice, in that pathogens and insects are not carried over to the next crop. In some areas of tropical Asia, however, straw is routinely plowed into the soil without serious trouble from diseases and pests. This return of rice straw to the soil is desirable for soil fertility, and some recent studies have proved its positive effect on yield (M. Isumunadji, personal communication). As stated in chapter 3, 60–70 percent of the total nitrogen taken up by rice plants comes from soil organic nitrogen, and the incorporation of straw enriches this source. But at the same time strong reduction caused by decomposition of straw may be a problem for rice plants at the early growth stage. Therefore, a careful study of both the positive and negative effects in cooperation with plant protection specialists is necessary before this practice is recommended to the farmer.

In aiming at higher yields, phosphorus would be a limiting factor in many areas of tropical Asia. Not only the availability but also the absolute content of phosphorus is critically low, even with present low yields (see chapters 5 and 9). Farmers in tropical Asia are aware of this. In the Kedah-Perlis Plain of West Malaysia bat guano, deposited in limestone caves, is taken and applied to the fields. In the Mekong delta of Vietnam a scaffold is set up as a habitat for bats, and their excreta is collected for manure. The amount of phosphorus thus applied is far less than that required in intensified rice cultivation, and chemical fertilizers have to be introduced. Amount and type of phosphate fertilizer must be determined for different types of soil.

In most soils potassium, calcium, and magnesium would be sufficient even in more intensive rice cultivation, as long as irrigation water can supply substantial amounts of these elements. In view of the large requirement for potassium, it should be applied to many soils. A study of the rate of release of soil potassium from different soil materials would help to establish potassic fertilizer recommendations.

In intensified rice cultivation some minor elements would become a limiting factor to yield increase in certain areas of tropical Asia. We think that many of the fan-terrace and plateau soils are liable to suffer from some kind of micronutrient deficiencies and also from silica deficiency. These soils, if double-cropped under irrigation, are also likely to show *akiochi* phenomena, because of their low, free iron-oxide content. Selection of fertilizers and management practices must be geared to avoid such difficulties.

(c) **Agricultural chemicals**

Intensification, involving heavy fertilizer use and/or double-cropping, would inevitably increase the incidence of pests and diseases.

Application of pesticides would become necessary for higher yields. Here the probable poisoning and subsequent decrease of the fish population must be considered. The fish catch in the paddy fields of tropical Asia has never been exactly weighed, but it is certainly large and has been an important source of protein. In view of this, the method and timing of the application of chemicals, and their fate in the soil and water, will have to be studied carefully.

(d) Mechanization

A soil scientist can make few suggestions in this field. The heavy clay soil in most big floodplains and deltas would pose difficulties for mechanization. The chief difficulties are the very hard dry consistency of the soil and the very sticky wet consistency of the same soil during the wet season. The soft ground of alluvial and deltaic plains also present difficulties. The physical properties of soils in relation to the workability of farm machinery have to be studied.

(e) Expansion of rice cultivation into virgin lands

In many countries of tropical Asia virgin lands for further expansion of rice cultivation are very rare and/or of very poor quality. The only large area left is along the coastal area of the insular part of Southeast Asia. In this area, potential and actual acid sulfate soils, and organic soils are predominant. Even the latter soils are often underlain by potential, acid sulfate soils. In fact, not much is known of the ecological condition of these tropical peat soils. Some studies have been carried out by Anderson (1964) in East Malaysia and by Polak (1943, 1951) and Driessen and Soepraptohardjo (1974) in Indonesia, but the effects of deforestation and drainage have not been studied in detail. It is well known that land subsidence and the decomposition of peat inevitably follow deforestation and drainage, but how rapidly? One area near Banjarmasin, South Kalimantan, was reclaimed in the 1930s for rice cultivation. The place is called Gambut, which means peat, but today there is no peat in the paddy fields of Gambut. Apparently all the peat has disappeared within the last forty years or so, leaving acidic clay soils. Fortunately, the area does not have a serious acid sulfate soil problem. One cannot, however, always rely on luck, and a thorough study of existing conditions and the possible effects of reclamation on the whole ecosystem is essential before utilization of these peaty swamp forest lands.

Another minor increase in paddy acreage could occur in hilly areas under rain-fed conditions (Takaya and Tomosugi, 1972a). Such expansion of acreage would contribute very little to total rice production, yet its adverse effect on water conservation and the run-off pattern would be considerable. Therefore, demarkation of good land suitable for intensified use and of land to be conserved, should be carried out urgently.

10.2 THE SOIL IN DIVERSIFICATION OF AGRICULTURE

One crop of rice should be taken for granted in the paddy fields of tropical Asia. Therefore, crop diversification means mainly introduction of upland crops other than rice in a double- or multiple-cropping system. As a matter of fact, in some areas of tropical Asia, double-cropping with an upland crop is practiced. For example, in the Chiang Mai basin in North Thailand, vegetables, pulses, and other commercial crops are fairly intensively cultivated as off-season crops, with irrigation. Cotton is cultivated, with irrigation, in Upper Burma. Without irrigation tobacco is cultivated in Kelantan, West Malaysia and the cultivation of peanuts is widespread in some parts of Burma, utilizing soil moisture remaining after rice cultivation. Pulses and rape are popular off-season crops in northern India and Bangladesh. In short, where water and soil tilth conditions are adequate, double- or multiple-cropping is practiced almost spontaneously. But if a major breakthrough in crop diversification is to occur, the problems of the vast areas of heavy deltaic soils and sandy terrace soils have to be solved.

(a) Physical properties

Upland crop cultivation makes different demands on soils. In the discussion of rice cultivation, soil physical properties have been almost totally ignored, but when upland crop cultivation is considered, these properties are even more important than chemical fertility. In heavy textured floodplain and delta soils, land preparation is very difficult, especially when an upland crop is to be cultivated after rice. In paddy fields the soil is usually elaborately puddled, which destroys the soil structure. After drying, the soil particles cohere to form very hard, big blocks, with many cracks, whereas when soil is wet it is too sticky to work with. Thus, the range of optimal moisture contents for good workability is very narrow. Insufficient land preparation under such conditions often leads to failure of seed germination. Even when seedlings are established, the low water-holding capacity of the soil due to the poor structure cannot assure an adequate water supply to the seedlings. Therefore, the most urgent need for successful crop diversification in deltaic soil areas is for a method of restoring soil structure quickly and inexpensively.

In sandy soils in fan-terrace and plateau areas, poor structure and low water-holding capacity are again the obstacles. Incorporation of manure and compost may be a solution, but acquisition of such materials is not always easy. Green manuring with adequate legumes is often advocated, but no successful examples have been reported.

(b) Chemical properties

All the major and minor nutrient problems stated in relation to intensified rice cultivation also apply to upland crop cultivation. In addition, the following should be noted:

    (1)   most upland crops cannot tolerate strong soil acidity
    (2)   phosphorus availability is much lower in upland conditions
    (3)   supply of nutrients by irrigation water is much less
    (4)   microbial nitrogen fixation is less, except in the case of legumes.

Therefore, both liming and larger amounts of major elements are required. In fan-terrace and plateau areas particularly, nutrient deficiencies, including those of micronutrients, would be more acute.

Here, too, application of manure and compost and green manuring may be recommended, but from the chemical point of view the after-effect of these organic manures on paddy field conditions, that is, strong reduction, should also be taken into account.

In small areas around big cities, for example, Bangkok, elaborate methods for continuous cultivation of such commercial crops as fruit and vegetables, is practiced, by building small polders and preparing raised beds in heavy textured deltaic soils. This requires quite high capital and labor inputs and is economically justified, for the time being, only where a large market is assured. But this example shows the possibility of turning heavy deltaic soils into a medium for upland crop cultivation. When continuously kept in upland condition, the maintenance of soil tilth seems to be comparatively easy even in these heavy textured soils. Therefore, the future development of crop diversification may follow this lead, with high rice productivity in a small area, enabling the rest of the land to be cultivated for upland crops.

References

Akuto, H. 1971. "What is Factor Analysis? Its Features and Problems." *Suri-Kagaku* [Mathematical Sciences], 94:14–20. (In Japanese.)

Anderson, J. A. R. 1964. "The Structure and Development of the Peat Swamps of Sarawak and Brunei." *Journal of Tropical Geography* 18:7–16.

Araragi, M.; and Tangcham, B. 1974. *Volatilized Loss of Soil Nitrogen and Microflora.* Mimeographed. Bangkok: Department of Agriculture.

Arnott, G. W. 1964. *A Tentative Guide to the Interpretation of Some Chemical Analyses of Malayan Soils.* Information Paper No. 246. Mimeographed. Kuala Lumpur: Division of Agriculture, Ministry of Agriculture and Co-operatives.

Asano, C. 1971. *Inshibunseki-ho Tsuron* [An Introduction to Internal and External Factor Analyses], Tokyo: Kyoritsu Shuppan. (In Japanese.)

Bloomfield, C.; and Coulter, J. K. 1973. "Genesis and Management of Acid Sulfate Soils" *Advances in Agronomy* 25:256–326.

Brammer, H. 1964. "An Outline of the Geology and Geomorphology of East Pakistan in Relation to Soil Development." *Pakistan Journal of Soil Science* 1:1–23.

———. 1971. "Coatings in Seasonally Flooded Soils." *Geoderma* 6:5–16.

——— and Hesse, P. R. 1970. "Some Unsolved Problems of Soil Science in East Pakistan." *Pakistan Journal of Soil Science* 6:31–44.

Breemen, N. van. 1976. *Genesis and Solution Chemistry of Acid Sulfate Soils in Thailand.* Wageningen: Centre for Agricultural Publishing and Documentation.

Brinkman, R. 1970. "Ferrolysis—A Hydromorphic Soil Forming Process" *Geoderma* 3:199–206.

Broeshart, H. 1971. "The Fate of Nitrogen Fertilizer in Flooded Rice Soils." *Nitrogen-15 in Soil-Plant Studies.* Vienna: IAEA.

Burma, Government of. 1972. *Report to the People, 1971–1972.* Rangoon.

Charoen, Prachak. 1974. "*Studies on Parent Material, Clay Minerals, and Fertility of Paddy Soils in Thailand.*" Doctoral dissertation, Kyoto University.

Cholitkul, W.; and Tyner, E. H. 1971. "Inorganic Phosphorus Fractions and Their Relation to Some Lowland Rice Soils of Thailand." *Proceedings of the International Symposium Soil Fertility Eval., New Delhi* 1:7–20.

Ciferri, R. Undated. *Lineamenti per Una Storia del Riso in Italia*, Ente Nazionale Risi, Quaderno No. 8.

Cooray, P. G. 1967. *An Introduction to the Geology of Ceylon.* Colombo: National Museums of Ceylon Publications.

Crocker, C. D. 1962. *Exploratory Survey of the Soils of Cambodia.* Mimeographed Phnom Penh: Royal Cambodian Government Soil Commission and

USAID Joint Publication.

Dost, H., ed. 1973. *Acid Sulphate Soils, Proceedings of the International Symposium, Volumes 1 and 2.* Wageningen: International Institute for Land Reclamation and Improvement. Publication No. 18.

Driessen, P. M.; and Soepraptohardjo, M. 1974. *Soils for Agricultural Expansion in Indonesia,* Bogor: Soil Research Institute.

Dudal, R. 1958. "Paddy Soils." *International Rice Commission News Letter* 7:19–27.

————. 1965. "Problem of Genesis and Classification of Paddy Soils." *Geography and Classification of Soils of Asia,* Moscow: Nauka. (In Russian.)

———— and Moormann, F. R. 1964. "Major Soils of Southeast Asia", *Journal of Tropical Geography,* 18:54–80.

———— and Soepraptohardjo, M. 1960. *Soil Classification in Indonesia,* Contribution of General Agricultural Research Station, No. 148, Bogor.

FAO. 1969. *Trade Yearbook, Vol. 23,* Rome.

————. 1974. *Production Yearbook, Vol. 28,* Rome.

————. 1974. *Trade Yearbook, Vol. 28,* Rome.

Fridland, V. M. 1964. "Tropical Rice Soils—On an Example of North Vietnam." *Reports on the Genesis and Geography of the Soils of Foreign Countries by the Soviet Geographers to the 8th International Congress of Soil Science.* Moscow: Nauka. (In Russian.)

Fukui, H. 1971. "Environmental Determinants Affecting the Potential Dissemination of High Yielding Varieties of Rice—A Case Study of the Chao Phraya River Basin." *Tonan Ajia Kenkyu* 9:348–374.

————. 1973. "Environmental Determinants Affecting the Potential Productivity of Rice—A Case Study of the Chao Phraya River Basin of Thailand." Doctoral dissertation, Kyoto University.

Furukawa, H.; Handawela, J.; Kyuma, K.; and Kawaguchi, K. 1976. "Chemical, Mineralogical, and Micromorphological Properties of Glaebules in Some Tropical Lowland Soils." *Tonan Ajia Kenkyu* 14: 365–388.

Furukawa, H.; and Kawaguchi, K. 1969. "Contribution of Organic Phosphorus to the Increase of Soluble Phosphorus upon Submergence." *Journal of Science of Soil and Manure, Japan* 40:141–148. (In Japanese.)

Hauck, R. D. 1971. "Quantitative Estimates of Nitrogen-Cycle Processes, Concepts and Review." *Nitrogen-15 in Soil-Plant Studies.* Vienna: IAEA.

Hayashi, C.; Higuchi, I.; and Komazawa, T. 1970. *Joho-shori to Tokei-suri* [Information processing and statistical mathematics]. Tokyo: Sangyo-tosho. (In Japanese.)

Imaizumi, K.; and Yoshida, S. 1958. "Edaphological Studies on Silicon Supplying Power of Paddy Soils." *Bulletin of the National Institute of Agricultural Sciences, Series B.* 8:261–304. (In Japanese.)

Inada, K. 1966. "Studies on the Bronzing Disease of Rice Plant in Ceylon, II Cause of the Occurrence of Bronzing." *Tropical Agriculturist* 125:31–46.

Ishizuka, Y.; and Tanaka, A. 1963. *Suito-no Eiyo-seiri* [Nutrio-physiology of rice plants], Tokyo: Yokendo. (In Japanese.)

Islam, M. A. 1955. *Reclamation of Kosh Soil of Chittagong Coastal Region.* Dacca: Department of Agriculture, Government of East Bengal.

Ismunadji, M.; Zulkarnaini, I.; and Miyake, M. 1975. *Sulphur Deficiency in*

Lowland Rice in Java, Contribution of Central Research Institute of Agriculture, No. 14, Bogor.

Kamoshita, Y. 1937. "Soil Types on Tsugaru Plain, Aomori Prefecture." *Journal of Science of Soil and Manure, Japan* 10:311–317. (In Japanese.)

Kanapathy, K. 1973. "Acid Swamp Soils, Their Occurrence, Formation, and Utilization." *Proceedings of the 2nd ASEAN Soil Conference, Jakarta* 1:247–254.

Kanno, I. 1956. "A Scheme for Soil Classification of Paddy Fields in Japan with Special Reference to Mineral Paddy Soils." *Bulletin of the Kyushu Agricultural Experimental Station* 4:261–273.

———. 1962. "A New Classification System of Rice Soils in Japan." *Pedologist* 6:2–10.

Karmanov, I. I. 1966. "Alteration of Tropical Soils under Cultivation" *Pochvovedenie* 1:36–48. (In Russian.)

Kawaguchi, K.; and Kyuma, K. 1969a. *Lowland Rice Soils in Thailand*. Kyoto: Center for Southeast Asian Studies, Kyoto University.

———. 1969b. *Lowland Rice Soils in Malaya*. Kyoto: Center for Southeast Asian Studies, Kyoto University.

———; and Matsuo, Y. 1954. "Determination of Free Iron Oxides in Soils" *Journal of Science of Soil and Manure, Japan* 25:31–35. (In Japanese.)

Kawasaki, I. 1953. *Natural Supplies of the Three Major Elements in Main Cultivated Soils in Japan*. Tokyo: Nippon Nogyo Kenkyusho. (In Japanese.)

Kayane, I. 1971. "Hydrological Regions in Monsoon Asia," in Yoshino, M., ed. *Water Balance of Monsoon Asia*. Tokyo: University of Tokyo Press.

Kevie, W. van der. 1972. "Morphology, Genesis, Occurrence and Agricultural Potential of Acid Sulphate Soils in Central Thailand." *Thai Journal of Agricultural Sciences* 5:165–182.

Kitagawa, Y.; Kyuma, K.; and Kawaguchi, K. 1973. "Clay Mineral Composition of Some Volcanogenous Soils in Indonesia and the Philippines." *Soil Science and Plant Nutrition* 19:147–159.

Kitano, Y. 1969. *Mizu-no Kagaku* [Water science]. Tokyo: Nippon Hoso Kyokai (NHK) Books, no. 92. (In Japanese.)

Kobayashi, J. 1958. "Chemical Studies on the Quality of the River Water in the Countries of Southeast Asia: Quality of Water in Thailand." *Nogaku-Kenkyu* 46:63–112. (In Japanese.)

———. 1961. "Mean Water Quality of the Japanese Rivers and Its Characteristics." *Nogaku-Kenkyu* 48:63–106. (In Japanese.)

Kobayashi, T. 1939. "Studies on Amelioration of Problem Soils in Polder Lands." *Ibaraki Agricultural Experimental Station, Special Report* 3:1–47. (In Japanese.)

Koenigs, F. F. R. 1950. "A 'Sawah' Profile near Bogor (Java)." *Transactions of the 4th International Congress of Soil Scientists, Amsterdam* 1:297–300.

Kojima, K.; Kittaka, A.; Yazawa, F.; and Shimoda, H. 1962. *Indonesiya-no Inasaku* [Rice Culture in Indonesia]. Tokyo: Japan FAO Association. (In Japanese.)

Komes, A. 1973. "The Reclamation of Some Problem Soils in Thailand." *Soils of the ASPAC Region; Part 5. Thailand*. Technical Bulletin No. 14, Food and Fertilizer Technology Center, Taipei.

Komoto, Y., *et al.* 1968. "Effect of Phosphate Fertilizers on the Rice in the Warm Region, Part 1." *Abstracts of the Annual Meeting of the Japanese Society of Soil Scientists* 14:94. (In Japanese.)

Koyama, T.; Chittana Chamek; and Natee Niamsrichand. 1973. *Nitrogen Application Technology for Tropical Rice as Determined by Field Experiments Using ^{15}N Tracer Technique.* Tokyo: Tropical Agricultural Research Center Technical Bulletin 3.

Kyuma, K. 1970. "Secondary Ferruginous Formations in Tropical Soils, Especially Concretions and Nodules" *Tonan Ajia Kenkyu* 7:571–581. (In Japanese.)

————. 1971. "Climate of South and Southeast Asia according to Thornthwaite's Classification Scheme." *Tonan Ajia Kenkyu* 9:136–158.

————. 1972a. "Numerical Classification of Climate, I. The Method and Its Application to the Climate of Japan." *Soil Science and Plant Nutrition* 18:155–167.

————. 1972b. "Numerical Classification of the Climate of South and Southeast Asia." *Tonan Ajia Kenkyu* 9:502–521.

————. 1973a. "Soil Water Regime of Rice Lands in South and Southeast Asia." *Tonan Ajia Kenkyu* 11:3–13.

————. 1973b. "A Method of Fertility Evaluation for Paddy Soils, II. Second Approximation: Evaluation of Four Independent Constituents of Soil Fertility." *Soil Science and Plant Nutrition* 19:11–18.

————. 1973c. "A Method of Fertility Evaluation for Paddy Soils, III. Third Approximation: Synthesis of Fertility Constituents for Soil Fertility Evaluation." *Soil Science and Plant Nutrition* 19:19–27.

————. 1976. *Paddy Soils in the Mekong Delta of Vietnam.* Discussion Paper No. 85. Mimeographed. Center for Southeast Asian Studies, Kyoto University.

————; Hattori, T.; and Kawaguchi, K. 1974. "Rice Cultivation and Environmental Conditions in the Mediterranean Countries, Part 2. Soils, Their Fertility and Mineralogy." *Soil Science and Plant Nutrition* 20:225–240.

————; and Kawaguchi, K. 1966. "Major Soils of Southeast Asia and the Classification of Soils under Rice Cultivation (Paddy Soils)." *Tonan Ajia Kenkyu* 4:290–312.

————. 1971. *Fertility Evaluation of Paddy Soils in South and Southeast Asia—First Approximation: Chemical Potentiality Rating.* Discussion Paper No. 34. Mimeographed. Center for Southeast Asian Studies, Kyoto University.

————. 1972a. *Fertility Evaluation of Paddy Soils in South and Southeast Asia—Second Approximation: Evaluation of Three Independent Constituents of Soil Fertility.* Discussion Paper No. 40, Mimeographed. Center for Southeast Asian Studies, Kyoto University.

————. 1972b. "An Approach to the Capability Classification of Paddy Soils in Relation to the Assessement of Their Agricultural Potential." *Transactions of the Second ASEAN Soil Conference, Jakarta* 1:283–294.

————. 1973. "A Method of Fertility Evaluation for Paddy Soil, I. First Approximation: Chemical Potentiality Grading." *Soil Science and Plant Nutrition* 19:1–9.

Kyuma, K.; Okagawa, N.; and Kawaguchi, K. 1974. "A Numerical Approach to the Classification of Alluvial Soil Materials." *Transactions of the 10th International Congress of Soil Scientists, Moscow* 6:543–551.

Löf, G. O. G., *et al.* 1966. *World Distribution of Solar Radiation.* Madison: University of Wisconsin. (Cited from Fukui, 1973.)

Matsuguchi, T.; and Tangcham, B. 1974. "Free-living Nitrogen Fixers and Acetylene Reduction in Tropical Paddy Field." *Transactions of the 10th International Congress of Soil Scientists, Moscow* 9:180–188.

Matsuo, Y. 1968. "Improvement of the Methods of Fertility Character Determination for the Paddy Soil Samples from Southeast Asia." *Tonan Ajia Kenkyu* 6:472–582. (In Japanese.)

Matsuzaka, Y. 1969. "Study on the Classification of Paddy Soils in Japan" *Bulletin of the National Institute of Agricultural Sciences, Series B.* 20:155–349. (In Japanese.)

Mitsuchi, M. 1974. "Pedogenic Characteristics of Paddy Soils and Their Significance in Soil Classification" *Bulletin of the National Institute of Agricultural Sciences, Series B.* 25:29–115. (In Japanese.)

————. 1975. "Permeability Series of Lowland Paddy Soils in Japan." *Japan Agricultural Research Quarterly* 9:28–33.

Miyake, I. 1973. *Shakaikagaku no Tameno Tokei-Pakkeiji* [Statistical package for social science] Tokyo: Toyoshinpo-sha. (In Japanese.)

Mohr, E. C. J.; and van Baren, F. A. 1954. *Tropical Soils.* The Hague: Mouton.

Money, D. C. 1965. *Climate, Soils and Vegetation,* London: University Tutorial Press.

Motomura, S. 1973. *Study on Advance in Rice Production by Soil Management: The Report of the Joint-Research Work between Thailand and Japan.* Mimeographed. Tropical Agricultural Research Center, Ministry of Agriculture and Forestry, Japan.

Murakami, H. 1965. *Studies on the Characteristics of Acid Sulfate Soils and Their Amelioration.* Doctoral dissertation, Kyoto University. (In Japanese.)

Netherlands Delta Development Team. 1973. *Soil Fertility,* Technical Note 15. Mimeographed. Mekong Committee, Bangkok.

————. 1974a. *Recommendations Concerning Agricultural Development with Improved Water Control in the Mekong Delta, Working Paper VA, Agriculture, Land Resources.* Mimeographed. Bangkok.

————. 1974b. *Recommendations Concerning Agricultural Development with Improved Water Control in the Mekong Delta, Working Paper IV, Hydrology.* Mimeographed. Bangkok.

Norrish, K.; and Chappel, B. W. 1967. "X-Ray Fluorescence Spectrography." in Zussman, J., ed., *Physical Method in Determinative Mineralogy.* New York: Academic Press.

Ojima, T.; and Kawaguchi, K. 1959. "The Reaction between Hardly Soluble Phosphate and Hydrogen Sulfide." *Bulletin of the Research Institute for Food Sciences, Kyoto University* 22:59–67. (In Japanese.)

Okuno, T.; Kume, H.; Haga, T.; and Yoshizawa, T. 1971. *Tahenryo Kaisekiho* [Multivariate statistical methods]. Tokyo: Nikka-Giren. (In Japanese.)

Otowa, M. 1969. "Development of Paddy Soil Classification in Japan." *Hokkaido Soils and Fertilizers News Letter* 63:14–21 Mimeographed. (In Japanese.)

Oyama, M. 1962. "A Classification System of Paddy Rice Field Soils Based on Their Diagnostic Horizons." *Bulletin of the National Institute of Agricultural Sciences, Series B.* 12:303–372. (In Japanese.)

Panabokke, C. R.; and Nagarajah, S. 1964. "The Fertility Characteristics of the Rice-Growing Soils of Ceylon." *Tropical Agriculturist* 120:1–27.

Patrick, W. H., Jr. 1964. "Extractable Iron and Phosphorus in a Submerged Soil at Controlled Redox Potentials." *Transactions of the 8th International Congress of Soil Scientists, Bucharest* 4:605–610.

————; and Mahapatra, I. C. 1968. "Transformation and Availability to Rice of Nitrogen and Phosphorus in Waterlogged Soils." *Advances in Agronomy* 20:323–359.

Pedologist's Subcommittee for Paddy Soil Classification. 1969. "On the 6th Tentative Classification Scheme." *Pedologist* 13:105–111. (In Japanese.)

Polak, B. 1943. *The Rawah Labkok (South-Priangan, Java), Investigation into the Composition of an Eutrophic, Topogenous Bog.* Chuo Noozi Sikenzyoo, Contribution No. 8, Bogor.

———— ; and Soepraptohardjo, M. 1951. *Pot and Field Experiments with Maize on Acid Forest Peat from Borneo.* Contribution of General Agricultural Research Station, No. 125, Bogor.

Ponnamperuma, F. N. 1972. "The Chemistry of Submerged Soils." *Advances in Agronomy* 24:29–96.

Research Group of Soil Microorganisms. 1966. "Nitrogen Fixation by Microorganisms." *Tsuchi to Biseibutsu* [Soil and microorganisms], Tokyo: Iwanami. (In Japanese.)

Rozanov, B. G.; and Rozanova, I. M. 1965. "Genesis of Degraded Paddy Soils in the Tropics." *Geography and Classification of Soils of Asia*, Moscow: Nauka. (In Russian.)

Sanchez, P. A. 1972. "Nitrogen Fertilization and Management in Tropical Rice." *North Carolina Agricultural Experiment Station Technical Bulletin*, 213. Raleigh, N. C.

Sarot Montrakun. 1964. *Agriculture and Soils of Thailand, Compilation of Soils Reports and Laboratory Analysis of the Soils in Thailand.* Bangkok: Ministry of Agriculture.

Seal, H. L. 1964. *Multivariate Statistical Analysis for Biologists.* London: Methuen & Co.

Shiga Agricultural Experiment Station and Ministry of Agriculture and Forestry. 1963. Mimeographed. *Report of the Designated Experiments* 3. (In Japanese.)

Sinanuwong, S.; and Takaya, Y. 1974a. "Saline Soils in Northeast Thailand— Their Possible Origin as Deduced from Field Evidence." *Tonan Ajia Kenkyu* 12:105–120.

————. 1974b. "Distribution of Saline Soils in the Khorat Basin of Thailand." *Tonan Ajia Kenkyu* 12:365–382.

Soil Research Institute. 1960. *Exploratory Soil Map of Jawa and Madura.* Bogor.

Soil Survey Staff. 1951. *Soil Survey Manual.* USDA Handbook No. 18, Washington, D.C.

———. 1960. *Soil Classification: A Comprehensive System: 7th Approximation.* USDA, Washington D.C.

———. 1967. *Soil Classification: A Comprehensive System: 7th Approximation, 1967 Supplement.* USDA, Washington, D.C.

Sokal, R. R.; and Sneath, P. H. A. 1963. *Principles of Numerical Taxonomy.* San Francisco: Freeman & Co.

Tadano, T. 1975. "Devices of Rice Roots to Tolerate High Iron Concentration in Growth Media." *Japan Agricultural Research Quarterly* 9:34–39.

Takahashi, E. 1961. *Nutrio-Physiological Studies of Silica to Crops.* Doctoral dissertation, Kyoto University (In Japanese.)

Takaya, Y. 1968. "Quaternary Outcrops in the Central Plain of Thailand." in Takimoto, K., ed., *Geology and Mineral Resources in Thailand and Malaya.* Kyoto: Center for Southeast Asian Studies, Kyoto University.

———. 1971. "Physiography of Rice Land in the Chao Phraya Basin." *Tonan Ajia Kenkyu* 9:375–397.

———. 1972. "Physiography of Rice Land in Peninsular Thailand." *Tonan Ajia Kenkyu* 10:422–432.

———; and Tomosugi, T. 1972a. "Rice Lands in the Upland Hill Regions of Northeast Thailand—A Remark on Rice Producing Forest" *Tonan Ajia Kenkyu* 10:77–85. (In Japanese.)

———. 1972b. "Three Categories of Rice Land in Northeast Thailand." *Ajia Keizai*, 13:66–72. (In Japanese.)

Takijima, Y. 1963. "Studies on Behaviour of the Growth Inhibiting Substances in Paddy Soils with Special Reference to the Occurrence of Root Damage in the Peaty Paddy Field." *Bulletin of the National Institute of Agricultural Sciences, Series B.* 13:117–252. (In Japanese.)

———; and Wijayaratna, H. M. S.; and Seneviratne, C. J. 1970. "Nutrient Deficiency and Physiological Disease of Lowland Rice in Ceylon, III. Effect of Silicate Fertilizers and Dolomite for Increasing Rice Yield." *Soil Science and Plant Nutrition* 16:11–23.

Tanaka, A.; and Yoshida, S. 1970. *Nutritional Disorders of the Rice Plant in Asia.* International Rice Research Institute, Technical Bulletin, 10, Manila.

———, *et al.* 1970. "A Note on the Nutritional Disorders of the Rice Plant in Java, Indonesia." *Tonan Ajia Kenkyu* 8:418–426.

Temple, K. L.; and Koehler, W. A. 1954. *West Virginia Engineering Experiment Station, Research Bulletin* 25. (Cited from Bloomfield, C.; and Coulter, J. K., 1973.)

Thornthwaite, C. W. 1948. "An Approach toward a Rational Classification of Climate." *The Geographical Review* 38:55–94.

Tommerup, E. C. 1934. *Transactions of the First Commission of the International Society of Soil Science* 155 (Paris). (Cited from Laatsch, W.: *Dynamik der Deutschen Acker- und Waldböden.* Dresden u. Leipzig: Verlag von Theodor Steinkopff, 1938, p. 81.)

Uchiyama, N. 1949. *Suiden-Dojo Keitai-ron* [Morphology of Paddy Soils], Tokyo: Chikyu-shuppansha. (In Japanese.)

Ueda, K., *et al.* 1975. "Effects of Continuous Application of Lime in Paddy Soils, Part 1. Relationships between Continuous Lime Application and Rice Yield." *Abstract of the Annual Meeting of the Japanese Society of Soil Scientists* 21, Part II: 59. (In Japanese.)

Uehara, G.; Nishina, M. S.; and Tsuji, G. Y. 1974. *The Composition of Mekong River Silt and Its Possible Role as a Source of Plant Nutrient in Delta Soils.* Mimeographed. Department of Agronomy and Soils Science, College of Tropical Agriculture, University of Hawaii.

Wada, H. 1966. "A Possible Classification of Paddy Soils according to the 7th Approximation." *Pedologist* 10: 141–145. (In Japanese.)

————; and Matsumoto, S. 1973. "Pedogenic Processes in Paddy Soils." *Pedologist* 17: 2–15.

Watabe, T. 1975. "Development of Rice Cultivation," in Ishii, Y., ed. *Tai-Koku—Hitotsu-no Inasaku-shakai* [Thailand—a rice-growing society], Tokyo: Sobunsha. (In Japanese.)

White House. 1967. *World Food Problem: A Report of the President's Science Advisory Committee.* Vol. II, Report of the Panel on the World Food Supply, Washington, D.C.

Yamada, N. 1959. "Some Aspects of the Physiology of Bronzing." *International Rice Commission Newsletter* 8: 11–16.

Yamane, I. 1975. *Nippon-no Shizen to Nogyo* [Nature and agriculture of Japan], Tokyo: Nobunkyo. (In Japanese.)

Yamazaki, K. 1960. "Studies on the Pedogenetic Classification of Paddy Soils in Japan." *Bulletin of the Toyama Prefectural Agricultural Experiment Station*, *Sp. Pub.* 1: 1–105. (In Japanese.)

Yoshida, S. 1961. "Natural Supply of Nutrients by Soil and River Water," in Konishi, C.; and Takahashi, J.; eds, *Dojo-Hiryo Koza* [Treatise on soils and fertilizers] 1. Tokyo: Asakura-Shoten. (In Japanese.) (Cited from Yamane, I, 1975.)

————; Navasero, S. A.; and Ramirez, E. 1969. "Effects of Silica and Nitrogen Supply on Some Leaf Characters of the Rice Plant." *Plant Soil* 31: 48–56.

Yoshida, T. 1971. "Soil Microbiology." *International Rice Research Institute Annual Report.*

————. 1972. "An Application of Studies on Soil Microorganisms to Rice Cultivation." *Hakko-Kyokai shi* [Japanese Association of Fermentation] 31: 1–8. (In Japanese.)

————; Roncal, R. A.; and Bautista, E. M. 1973. "Atmospheric Nitrogen Fixation by Photosynthetic Microorganisms in a Submerged Philippine Soil." *Soil Science and Plant Nutrition* 19: 117–123.

————; and Padre, B. C. Jr. 1974. "Nitrification and Denitrification in Submerged Maahas Clay Soil." *Soil Science and Plant Nutrition* 20: 241–247.

Appendix 1
List of Samples

(1-location; 2-landform and parent material; 3-local soil classification)

BANGLADESH

Bd-1   1   Sarolia, P. S. Tejgaon, Dacca District
          2   Upper part of very gently undulating land on subrecent Brahmaputra alluvium
          3   Jalkundi-Naraibag series

Bd-2   1   Sarolia, P. S. Tejgaon, Dacca District; 200 meters south southeast of the Bd-1 site
          2   Lower part of very gently undulating land on subrecent Brahmaputra alluvium
          3   Siddhirganj series

Bd-3   1   Sarolia, P. S. Tejgaon, Dacca District; 400 meters west of the Bd-1
          2   Upper part of very gently undulating land on subrecent Brahmaputra alluvium
          3   Jalkundi series

Bd-4   1   Hatimara, P. S. Savar, Dacca District
          2   Level land on old alluvium of Madhupur Tract
          3   Chandra series?

Bd-5   1   Auspara, P. S. Joydebpur, Dacca District; 14.5 miles north of Dacca, 300 meters west of the road
          2   Middle part of gently undulating land on old alluvium of Madhupur Tract
          3   Demra series

Bd-6   1   Badshartek, P. S. Tejgaon, Dacca District; 11.2 miles north of Dacca
          2   Level land on old alluvium of Madhupur Tract
          3   Chhiata series

Bd-7   1   Azampur, P. S. Tejgaon, Dacca District; 11.2 miles north of Dacca, the other side of the road from Bd-6
          2   Level bottom of a shallow narrow valley on old alluvium of Madhupur Tract
          3   Kalma series

Bd-8   1   Holyvta, P. S. Dhamrai, Dacca District; 24.5 miles from Dacca on Manikganj Road, 0.25 mile north of the road
          2   Lower part of gently undulating land on recent Jamuna alluvium
          3   Sabhar Bazar series

Bd-9   1   Jaleshwar, P. S. Sabhar, Dacca District; 16.8 miles from Dacca, 30 meters west of the road

    2   Level land on old alluvium of Madhupur Tract

    3   Noadda series

Bd-10 1   Hemayetpur, P. S. Sabhar, Dacca District; 12.8 miles west of Dacca on Sabhar road, 100 meters north of the road

    2   Upper part of gently undulating land on recent Jamuna alluvium

    3   Sanatola series

Bd-11 1   Chandulia, P. S. Sabhar, Dacca District

    2   Middle part of gently undulating land on recent Jamuna alluvium

    3   Dhamrai series

Bd-12 1   Daschira, P. S. Sibalay, Dacca District; 50.8 miles from Dacca on Manikganj road, 0.33 mile south of the road

    2   Lower part of gently undulating land on subrecent Gangetic alluvium.

    3   Rathuria series?

Bd-13 1   Daschira, P. S. Sibalay, Dacca District; 300 meters south of Bd-12

    2   Upper part of gently undulating land on subrecent Gangetic alluvium

    3   Sara series

Bd-14 1   Between Bd-12 and Bd-13

    2   Middle part of gently undulating land on subrecent Gangetic alluvium

Bd-15 1   Turag, P. S. Sabhar, Dacca District; 1 mile west of Mirpur Bridge (on the Turag River), 0.33 mile south of the road

    2   Upper part of very gently undulating land on recent Jamuna alluvium

    3   Pagla series?

Bd-16 1   Badda, P. S. Tejgaon, Dacca District; byde land between Badda and Merul

    2   Lower part of sideslope of a byde on local alluvial material of Madhupur Tract

    3   Kilgaon Series

Bd-17 1   Badda, P. S. Tejgaon, Dacca District; byde land between Badda and Merul, about 5 feet above the Bd-16

    2   Middle part of side slope of a byde on local alluvium material of Madhupur Tract

Bd-18 1   South of Madhupur 49.5 miles to Mymensingh on Dacca-Mymensingh road

    2   Level land on subrecent Brahmaputra alluvium

Bd-19 1   Kaitkai, P. S. Madhupur, Mymensingh District; 0.5 mile from Madhupur junction on Madhupur-Jawalpur road

    2   Level levee land on subrecent Brahmaputra alluvium

    3   Lokdeo series?

Bd-20 1   Gobrakura, P. S. Haluaghat, Mymensingh District; 2 miles north of Haluaghat, 300 meters east of the road

    2   Upper to middle part of gently sloping land on subrecent piedmont plain sediment

	3	Jhinaigati series
Bd-21	1	Barakalhar, P. S. Gouripur, Mymensingh District
	2	Lower part of very gently undulating land on subrecent Brahmaputra alluvium
Bd-22	1	Gavishimul, P. S. Gouripur, Mymensingh District
	2	Level land on subrecent Brahmaputra alluvium
	3	Silmandy series?
Bd-23	1	Dewannagar, P. S. Hathazari, Chittagong District; about 1 mile south of Hathazari, check plot of the experimental field of the Soil Fertility Institute
	2	Middle part of gently sloping land on subrecent piedmont plain sediment
	3	Fatiabad-Pahartali series?
Bd-24	1	Noapara, P. S. Raozan, Chittagong District; 11 miles from Chittagong on Kapthai road, 100 meters north of the road
	2	Lower part of very gently undulating land on recent Karnaphli alluvium
	3	Guman-Mardan series
Bd-25	1	Patia, P. S. Patia, Chittagong District
	2	Level land on subrecent river (Karnaphli, etc.) alluvium
	3	Hathazari series?
Bd-26	1	127 miles from Dacca on Chittagong road, neat Feni
	2	Middle part of gently sloping land on subrecent Piedmont plain sediment
Bd-27	1	Bhozpara, P. S. Laksham, Comilla District; 2 miles east of Laksham, 300 meters south of the road
	2	Upper to middle part of very gently undulating land on subrecent Brahmaputra alluvium
	3	Dhamti series
Bd-28	1	Ashruffpur, P. S. Sadar, Comilla District; 0.5 mile west of Comilla Airport
	2	Shallow basin site on subrecent piedmont outwash material
	3	Bharella series?
Bd-29	1	Gumta, P. S. Muradnagar, Comilla District; 11.1 miles east of Daudkandi
	2	Lower part of very gently undulating land on subrecent Brahmaputra alluvium
	3	Naraibag series?
Bd-30	1	Gupchar, P. S. Daukandi, Comilla District; 4 miles east of Daudkandi, about 2 miles north of the Comilla road, on a char between Gumti and a minor river near Gouripur village
	2	Middle part of gently undulating land on recent Gumti alluvium
	3	Gumti series?
Bd-31	1	Tetri, P. S. Sadar, Sylhet District; 3 miles from Sylhet on Moulvi Bazar Rd., 200 meters east of the road
	2	Lower part of gently undulating land on subrecent Surma alluvium

       3   Phagu series

Bd-32 1   Tetri, P. S. Sadar, Sylhet District; 3 miles from Sylhet on Moulvi Bazar road, 200 meters west of the road

       2   Upper to middle part of gently undulating land on subrecent Surma alluvium

       3   Goyainghat series

Bd-33 1   Mirzapur, P. S. Kulaura, Sylhet District; 200 meters west of Brahman Bazar (about 5 miles west of Kulaura), 100 meters north of the road

       2   Middle part of very gently sloping land on subrecent piedmont plain sediment

       3   Pritingpasa series

Bd-34 1   Babnaban, P. S. Kulaura, Sylhet District; 0.33 mile east of railroad 1 mile south of Kulaura

       2   Lower part of a very gently sloping land on subrecent piedmont plain sediment

       3   Manu series

Bd-35 1   Chawk Ramendapur (Radhanagar), P. S. Kotwali, Pabna District; 2 miles west of Pabna

       2   Level land on subrecent Gangetic alluvium

       3   Ishurdi series

Bd-36 1   N.D.T.I. Farm at Natore

       2   Upper part of gently undulating land on subrecent Gangetic alluvium

Bd-37 1   N.D.T.I. Farm at Gaibanda

       2   Level land on recent Telsta alluvium

Bd-38 1   Monthana, P. S. Kotwali, Rangpur District; 6 miles west of Rangpur

       2   Middle part of gently sloping land on recent Teesta alluvium

       3   Kaunia series?

Bd-39 1   Paglapir, P. S. Kotwali, Rangpur District; 8 miles west of Rangpur

       2   Upper part of gently undulating land on recent Teesta alluvium

       3   Pirgachha series?

Bd-40 1   Dalua, P. S. Birganj, Dinajpur District; 8 miles south of Thakulgaon

       2   Middle part of gently sloping land on subrecent piedmont plain sediment

Bd-41 1   Haribur Jot Samas, P. S. Boda, Dinajpur District; 6 miles south of Panchagar

       2   Level land on subrecent piedmont plain sediment

Bd-42 1   Azrampur, P. S. Pirganj, Rangpur District; 26.5 miles south of Rangpur

       2   Level land on old alluvium of Barind Tract

       3   Chandra series?

Bd-43 1   Khedmatpur, P. S. Pirganj, Rangpur District; 27.5 miles south of Rangpur

      2 Upper part of gently sloping land on odd alluvium of Barind Tract

      3 Belabo series

Bd-44 1 Lakshmipur, P. S. Kotwali, Faridpur District; about 9 miles west of Faridpur on Jessore road

      2 Lower part of gently undulating land on subrecent Gangetic alluvium

      3 Eral series?

Bd-45 1 N.D.T.I. Farm, P. S. Daulatpur, Khulna District; west of the staff quarters

      2 Lower part of gently undulating land on subrecent Gangetic alluvium.

Bd-46 1 Golna, P. S. Dumuria, Khulna District; 1 mile from Thana H. Q.

      2 Upper part of very gently undulating land on subrecent Gangetic alluvium

      3 Amadi series

Bd-47 1 0.5 mile north of Haripar road junction on Daulatpur-Satkhira road

      2 Depressed level land on recent brackish alluvium

Bd-48 1 Raimahal, P. S. Daulatpur, Khulna District; 0.5 mile south of the west edge of a village located west of Khulna Girls' College, about 100 meters east of the road

      2 Extensive swampy land on recent brackish alluvium

Bd-49 1 Sundarpur, P. S. Morirampur, Jessore District; 14 miles south of Jessore on Kesabpur road

      2 Middle part of gently undulating land on subrecent Gangetic alluvium

      3 Kobadak series?

Bd-50 1 Block II, Plot 8, WAPDA Amla Farm, P. S. Mirpur, Kushtia District

      2 Lower part of gently undulating land on subrecent Gangetic alluvium

      3 Eral series

Bd-51 1 Block VII, plot 2, WAPDA Amla Farm, P. S. Mirpur, Kushtia District

      2 Upper part of gently undulating land on subrecent Gangetic alluvium

      3 Amla series

Bd-52 1 Block IX, plot 11, WAPDA Amla Farm, P. S. Mirpur, Kushtia District

      2 Shallow basin in gently undulating land on subrecent Gangetic alluvium

      3 Fatki series

Bd-53 1 Shibpur, P. S. Baraigram, Rajshahi District; Site 4, photo B14-12

      2 Lower part of very gently sloping abandoned levee on subrecent Gangetic alluvium

      3 Ghior series

BURMA

B-1	1	Seed farm, Thayaung Chaung, Bassein, Irrawaddy
	2	Level land on recent alluvium
B-2	1	Kan Ywa, Bassein, Irrawaddy
	2	Level land on recent alluvium
B-3	1	Ywatha Gyi, Myaungmya, Irrawaddy
	2	Lower part of gently sloping land on recent alluvium
B-4	1	Kwin Yaa, Bassein, Irrawaddy
	2	Level land on recent alluvium
B-5	1	Byaik Gyi, Tyaung Gun, Irrawaddy
	2	Depressed level land on recent alluvium
B-6	1	Ywa Thit Kone, Bassein, Irrawaddy
	2	Upper part of gently undulating land on old alluvium
B-7	1	Shan Ga Lay Kyun, Mandalay (nonpaddy land)
	2	Kaing-land, sand bar in the Irrawaddy
B-8	1	Yenatha Tank, Yenatha, Mandalay
	2	Upper part of gently sloping land on subrecent alluvium
B-9	1	Tahudtaw, Pathein Gyi, Mandalay
	2	Middle part of gently undulating land on subrecent alluvium
B-10	1	Thayetkan, Madaya, Mandalay
	2	Depressed level land on recent alluvium
B-11	1	Gwegon Experiment Station, Shwebo, Sagaing
	2	Level land on subrecent alluvium
B-12	1	Kyaukse Central Agricultural Farm, Kyaukse, Mandalay
	2	Level land on subrecent alluvium
B-13	1	Hlwepauk, Myitha, Kyaukse, Mandalay
	2	Middle part of gently sloping land on subrecent alluvium
B-14	1	Tatkon Central Agricultural Farm, Tatkon, Mandalay (nonpaddy land)
	2	Level land on recent sandy alluvium
B-15	1	Agricultural Research Institute, Yezin, Mandalay
	2	Upper part of sloping land on subrecent alluvium
B-16	1	Agricultural Research Institute, Yezin, Mandalay
	2	Lower part of sloping land on subrecent alluvium

CAMBODIA

Ca-1	1	Koy Meng, Mongkol Borey, Battambang, right bank of the Mongkol Borey River
	2	Level land on recent lacustrine alluvium
Ca-2	1	Koy Meng, Mongkol Borey, Battambang, left bank of the Mongkol Borey River
	2	Level land on recent lacustrine alluvium
Ca-3	1	Centre Technique Agricole, Field No. 3, Tuol Samrong, Battambang
	2	Lower part of gently sloping land on old alluvium
Ca-4	1	Centre Technique Agricole, Field No. 11, Tuol Samrong, Battam-

bang

	2	Upper part of gently sloping land on old alluvium
Ca-5	1	5 km northeast along National Road from Komgpong Luong Ferry and 200 meters north of road, Kompong Cham Province
	2	Level land on subrecent alluvium
	3	Lacustrine alluvial soils
Ca-6	1	Khum Khvav, Komgpong Cham Province (16 km from Kompong Cham).
	2	Lower part of terraced sloping land on residuum from basalt
	3	Regurs, basaltic
Ca-7	1	Thnal Dach, Siem Reap Province, 7 km south of the highway on Kongpom Kleang road, 50 meters west of the road, 50 meters north of the freshwater mangrove swamp.
	2	Depressed level land on recent lacustrine alluvium
	3	Lacustrine alluvial soils
Ca-8	1	Bat Meas, Khum Samrong, Srok Sot Nikum, 8 km north of Damdek on Kok Koy road, 100 meters east of the road.
	2	Level land on subrecent alluvium.
	3	Gray hydromorphics.
Ca-9	1	25 km north on Samrong road from the Kralanh junction, 50 meters west of the road.
	2	Upper part of undulating land on old alluvium
	3	Cultural hydromorphics
Ca-10	1	1 km west of Kralanh, 200 meters east of Stung Srey Bridge, 60 meters south of the road
	2	Level land on recent alluvium
	3	Alluvial soils
Ca-11	1	A little less than 18 km west of Mongkol Borei (about 9 km west of Sisophon) on Poipet road, 100 meters north of the road
	2	Level land on old alluvium
	3	Plinthite podzol
Ca-12	1	28 km southeast of Battambang on Pursat road, 150 meters southeast of the road
	2	Lower part of gently sloping land on old alluvium
	3	Brown hydromorphics
Ca-13	1	Phum Prey Koki, Khum Churn Mtes, Srok Svey Teap, Khet Svey Rieng, 155 km from Phnom Penh on Saigon highway, 60 meters south of the road
	2	Level land on recent alluvium (brackish?)
	3	Alumisols
Ca-14	1	Ponsaing, Srok Prey Brabas, Takeo; 17.5 km from the Takeo road (from 11.5 km south of Chanbak); 100 meters north of the road
	2	Lower part of sloping land on old alluvium
	3	Alluvial soils
Ca-15	1	1 km north of Chanbak, Takeo; 100 meters west of the road
	2	Upper part of gently sloping land on old alluvium

	3	Cultural hydromorphics
Ca-16	1	Khum Banteay Dek; Srok Kean Sray Khet Kandal; 28 km southeast of Phnom Penh on Sray Rieng road
	2	Middle part of gently sloping land on recent alluvium; levee-backswamp transition
	3	Brown alluvial soils

INDIA

I-1	1	Kharka, Akharpur, Kanpur, Uttar Pradesh; 4 miles west of Muriapur Junction
	2	Level land on subrecent alluvium
	3	Kanpur Type IV
I-2	1	0.25 mile north along the canal passing Muriapur on the Bhognipur-Ghatampur road, Kanpur District, Uttar Pradesh
	2	Level land on subrecent alluvium
	3	Kanpur Type IV or V
I-3	1	Hamirpur Road Railway Station, Baripal, Kanpur, Uttar Pradesh
	2	Middle part of undulating land on subrecent alluvium
I-4	1	Pipra, Naugarh, Basti, Uttar Pradesh
	2	Middle part of very gently sloping land on subrecent alluvium Terai area
I-5	1	Patharba, Padrauna, Deoria, Uttar Pradesh
	2	Level land on calcareous subrecent alluvium (Gandak River?)
	3	Locally known as Bhat soil
I-6	1	Phagnia, Chandauli, Varanasi, Uttar Pradesh; 3 miles north of Chandauli
	2	Middle part of very gently sloping land on subrecent alluvium (Ganges)
	3	Locally known as Dhankar
I-7	1	Not recorded
	2	Depressed level land on recent alluvium
	3	Karail (local name for a Grumusolic Alluvial soil)
I-8	1	Bikramganj Irrigation Research Substation, Shahabad, Bihar; 36 miles from Arrah
	2	Level land on old alluvium upland position
I-9	1	Agraitula, Ramgarh, Sasaram, Shahabad, Bihar; 5 miles north of Sasaram
	2	Level land on old alluvium medium upland position
I-10	1	Moker, Sasaram, Shahabad, Bihar; 3 miles north of Sasaram
	2	Level land on old alluvium medium lowland position
I-11	1	Panchlakh, Khijirsarai, Gaya, Bihar
	2	Middle part of gently undulating land on old alluvium
I-12	1	Kuchasin Gaya Muffassil, Gaya, Bihar
	2	Level land on old alluvium
I-13	1	Hasanchak, Harnaut, Pabna, Bihar
	2	Level land on old alluvium

I-14   1   Kharuara, Harnaut, Pabna, Bihar
        2   Depressed level land on old alluvium
I-15   1   Narsara, Darbhanga, Darbhanga, Bihar
        2   Level land on subrecent alluvium (Bhagmati)
I-16   1   Sahuri, Kalyanpur, Darbhanga, Bihar
        2   Level land on calcareous subrecent alluvium (Burhi Ghandak)
I-17   1   Khoksibagh, Purnea, Purnea, Bihar
        2   Depressed level land on recent alluvium (Kosi)
I-18   1   Damka, Purnea, Purnea, Bihar
        2   Level land on recent alluvium (Kosi)
I-19   1   Maria, Purnea, Purnea, Bihar
        2   Level land on recent alluvium (Kosi), upland
I-20   1   Bangaon, Rajan, Bhagalpur, Bihar: 16 miles south of Bhagalpur
        2   Middle part of undulating land on old alluvium
I-21   1   Bangaon, Rajaon, Bhagalpur, Bihar; 18 miles south of Bhagalpur
        2   Middle part of gently sloping land on old alluvium
I-22   1   Akshaynagar, Kagdip, 24 Parganas, West Bengal
        2   Depressed low land on recent alluvium
I-23   1   21 miles from Calcutta towards Kagdip, West Bengal
        2   Level land on recent alluvium
I-24   1   Burdwan Seed Farm, Burdwan, West Bengal
        2   Middle part of gently sloping land on subrecent alluvium
            (Damodar)
I-25   1   Chinsula Rice Research Station Farm, Hooghly; control plots
            ($(NH_4)_2SO_4$ experiment for 18 years)
        2   Depressed level land on subrecent alluvium (Ganges)
I-26   1   Durgabati, Barsul, Burdwan, West Bengal
        2   Level land on subrecent alluvium
I-27   1   Bhubaneswar Agricultural Experiment Station, Orissa University
            of Agriculture and Technology, Bhubaneswar, Orissa
        2   Upper part of gently sloping land on local alluvium, derived from
            laterized sandstone
I-28   1   Central Rice Research Institute Farm, Cuttack, Orissa; Block J3C
        2   Level land on subrecent alluvium (Mahanadi)
I-29   1   Central Rice Research Institute Farm, Cuttack, Orissa; Block N
        2   Level land on subrecent alluvium (Mahanadi)
I-30   1   Kujang Farm, 50 miles east of Cuttack on Paradeep road, Orissa
        2   Level land on recent levee alluvium (Mahanadi)
I-31   1   Balijhara, Paradeep, Orissa, 3 km west of the port
        2   Depressed level land on recent alluvium (marine)
I-32   1   Barikel, Padampur, Sambalpur, Orissa; adjacent to Barikel Seed
            Farm
        2   Upper part of sloping land on local alluvium derived from gneissic
            rocks
I-33   1   Barikel Seed Farm, Padampur, Sambalpur, Orissa
        2   Basin in undulating land on local alluvium derived from gneissic

		rocks
I-34	1	Labhandi Farm, IADP, Raipur, Madhya Pradesh
	2	Basin in gently undulating land on local alluvium derived from shale (?)
	3	Locally known as Kanhar
I-35	1	Labhandi Farm, Rice Research Station, Raipur, Madhya Pradesh
	2	Middle part of very gently sloping land on local alluvium derived from shale (?)
	3	Locally known as Matasi
I-36	1	Labhandi Farm, Rice Research Station, Raipur, Madhya Pradesh
	2	Lower part of very gently sloping land on local alluvium derived from shale (?)
	3	Locally known as Dorsa
I-38	1	2 miles northeast of Balugaon, Orissa
	2	Level land on recent alluvium
I-39	1	Kashpur Farm, Saline Research Substation, Khallikot, Ganjam, Orissa
	2	Depressed level land on recent alluvium (marine)
I-40	1	Ragolu, Srikakulam, Srikakulam, Andhra Pradesh
	2	Lower part of very gently undulating land on subrecent alluvium (Nagavali)
I-41	1	Tardiput, Jeypur, Koraput, Orissa
	2	Middle part of gently sloping land on local alluvium derived from charnockite
I-42	1	Lamtaput, Machkund, Koraput, Orissa
	2	Lower part of gently sloping land on local alluvium derived from charnockite
I-43	1	Kasimkota, Kasimkota, Anakapalle, Andhra Pradesh
	2	Upper part of undulating land on local alluvium derived from gneiss
I-44	1	Yallamanchily, Yallamanchily, Vishakhapatnam, Andhra Pradesh
	2	Level land on subrecent alluvium (Sarada)
	3	Red sandy loam
I-45	1	Peddapuram, Peddapuram, East Godavari, Andhra Pradesh
	2	Level land on recent alluvium (Yeleru)
I-46	1	Jaggammagaripet, Samalkot, East Godavari, Andhra Pradesh
	2	Level land on recent alluvium, underlain by old alluvium
I-47	1	State Seed Multiplication and Demonstration Farm, Samalkot, East Godavari, Andhra Pradesh
	2	Level land on subrecent alluvium
I-48	1	Chollongi, Kakinada, East Godavari, Andhra Pradesh
	2	Level land on recent coastal sandy alluvium
I-49	1	Chollongi, Kakinada, East Godavari, Andhra Pradesh
	2	Depressed level land on recent coastal sandy alluvium
I-50	1	Polekurru, Kakinada, East Godavari, Andhra Pradesh
	2	Level land on recent coastal sandy alluvium

I-51	1	Neelapalli, Kakinada, East Godavari, Andhra Pradesh
	2	Level land on recent alluvium (lagoonal?)
I-52	1	Bicarole, Ramachandrapuram, East Godavari, Andhra Pradesh
	2	Middle part of gently undulating land on old alluvium
I-53	1	Pandalapaka, Ramachandrapuram, East Godavari, Andhra Pradesh
	2	Depressed level land on recent alluvium
I-54	1	Kotapeta, Kotapeta, East Godavari, Andhra Pradesh
	2	Level land on recent alluvium
	3	Black alluvium
I-55	1	Coconut Research Station Farm, Ambajipeta, Amalapuram, East Godavari, Andhra Pradesh
	2	Level land on recent alluvium
	3	Black alluvium
I-56	1	Mukkamala, Tanuku, West Godavari, Andhra Pradesh
	2	Level land on recent alluvium
	3	Black alluvium
I-57	1	Agricultural Research Station Farm, Tenali, Guntur; Andhra Pradesh. Field 2
	2	Level land on recent alluvium (marine)
I-58	1	Votluru Railway Crossing, west of Eloru, Andhra Pradesh
	2	Lower part of gently undulating land on old alluvium
I-59	1	Anumanchipalli, Nandigama, Krishna, Andhra Pradesh
	2	Lower part of undulating land on local calcareous alluvium derived from gneissic rocks
I-60	1	Edulabad, Hyderabad East, Hyderabad, Andhra Pradesh
	2	Middle part of undulating to rolling land on local alluvium derived from granite
	3	Red loam, locally known as Chalka
I-61	1	6 miles west of Hyderabad, Hyderabad West, Hyderabad, Andhra Pradesh
	2	Lower part of undulating land on local alluvium derived from gneiss
I-62	1	Agricultural Research Institute, Rajandranagar, Hyderabad, Andhra Pradesh; Farm 118–120
	2	Basin site of undulating land on recent alluvium (Musi)
I-63	1	Agricultural Research Institute, Rajandranagar, Hyderabad, Andhra Pradesh
	2	Middle part of sloping (NE) land on local alluvium derived from granite
	3	Chalka turned paddy soil
I-65	1	Rice Research Station Farm, Tenali, Guntur, Andhra Pradesh; Field 2
	2	Level land on subrecent alluvium
I-66	1	Chebrole, Tenali, Guntur, Andhra Pradesh
	2	Level land on subrecent alluvium
I-67	1	Masori, Kavali, Nellore, Andhra Pradesh

	2	Level land on old alluvium
I-68	1	Allur, Buchireddipalem, Nellore, Andhra Pradesh
	2	Level land on recent alluvium (marine)
I-69	1	North-Mopur, Buchireddipalem, Nellore, Andhra Pradesh; 16 miles from Nellore on Allur road
	2	Level land on recent coastal sandy alluvium
I-70	1	Gandawaram, Kovuru, Nellore, Andhra Pradesh; 6 miles north of Nellore on Kavali road
	2	Level land on subrecent alluvium
I-71	1	Kovuru, Kovuru, Nellore, Andhra Pradesh
	2	Level land on subrecent alluvium
I-72	1	Kakatur, Venkatachalam, Nellore, Andhra Pradesh
	2	Level land on old alluvium
I-73	1	Regional Rice Research Station Farm, Nellore, Andhra Pradesh
	2	Level land on recent alluvium
I-74	1	Padagachcheri, Papanasam, Tanjavur, Madras
	2	Level land on subrecent alluvium
I-75	1	Saliamangalam, Ammapettai, Tanjavur, Madras
	2	Level land on old alluvium

INDONESIA

In-1	1	Kedung Halang, Kedung Halang, Bogor, West Java
	2	Upper part of sloping land (in rolling country) on old volcanic fan deposits
	3	Reddish brown Latosol
In-2	1	Tjiriung, Thibinong, Bogor, West Java
	2	Lower part of undulating land on old volcanic fan deposits
	3	Yellowish red Latosol
In-3	1	Field of Akademi Department Pertanian, Bendungan, Tjiawi, Bogor, West Java
	2	Upper part of sloping land on foot slope colluvium of Mt. Pangrango
	3	Brown Latosol
In-4	1	Block Kramat, Afd. Sukamantri, Kotabatu, Bogor, West Java
	2	Middle part of sloping land on footslope colluvium of Mt. Salak
	3	Andosol
In-5	1	Kebun Besar, Batu Tjeper, Tangerang, West Java
	2	Lower part of very gently undulating land on old alluvium
	3	Red Latosol
In-6	1	Kebun Pertjobaan, Singa Merta, Tjiruas, Serang, West Java
	2	Level, recent alluvium
	3	Gray hydromorphic and Planosol
In-7	1	Petung, Sentul, Kragilan, Serang, West Java
	2	Level, moderately dissected old erosional terrace composed of acid tuff
	3	Planosol
In-8	1	Pasir Gombong, Lamahabang, Bekasi, West Java

	2	Level land on old alluvium
	3	Red-yellow Podzolic
In-9	1	Palawad, Karawang, Karawang, West Java
	2	Level land on recent alluvium
	3	Alluvial
In-10	1	Production Farm of Lembaga Sanghiang Seri, Sukamandi, Subang, West Java
	2	Lowest part of gently undulating land on old alluvium
	3	Red-yellow podzolic-Gray Hydromorphic
In-11	1	Block G2, Kebun Pertjobaab LPPP, Pusakanegara, Subang, West Java
	2	Level land on subrecent alluvium
	3	Gray brown alluvial
In-12	1	Sudikampiran, Slieg, Indramaju, West Java
	2	Upper part of level to very gently undulating land on recent alluvium
	3	Gray brown alluvial
In-13	1	Sampora, Tjirimus, Kuningan, West Java
	2	Middle part of sloping land on footslope colluvium of Mt. Tjirumai
	3	Brown Latosol
In-14	1	Pamojanan, Ketapang, Pameungpeuk, Bandung, West Java
	2	Level to very gently sloping land on recent alluvium
	3	Dark brown alluvial
In-16	1	Warung Kaweri, Tjipagran, Tjimahi, Bandung, West Java
	2	Middle part of moderately steep slope with terracing on footslope colluvium of Mt. Burangrang
	3	Brown Latosol
In-17	1	Farm of Persahaan Pertanian, Tjihea, Darmaga, Tjirandjang-girang, Bodjongpitjung, Tjiandjur
	2	Lower part of gently undulating land on old alluvium
	3	Grumusol
In-18	1	Medini, Undaan, Kudus, Central Java
	2	Level land on subrecent alluvium
	3	Alluvial
In-19	1	Majong Lor, Majong, Djepara, Central Java
	2	Gently sloping land on footslope colluvium of Mt. Muria
	3	Dark brown or brown Mediterranean
In-20	1	Katonsari, Demak, Demak, Central Java
	2	Level land on recent alluvium
	3	Grumusol
In-21	1	Kartohardjo, Buaran, Pekalongan, Central Java
	2	Level land on subrecent alluvium
	3	Alluvial
In-22	1	Sirandu, Pemalang, Pemalang, Central Java
	2	Level land on subrecent alluvium
	3	Alluvial

In-23   1   Seed Farm, Bulakamba, Bulakamba, Brebes, Central Java
        2   Level land on subrecent alluvium
        3   Alluvial
In-24   1   Bodjong, Purbolingo, Purbolingo, Central Java
        2   Level to very gently sloping land on subrecent alluvium
        3   Alluvial
In-25   1   Ladjer, (No. 34), Ambil, Kebumen, Central Java
        2   Level to very gently sloping land on subrecent alluvium
        3   Alluvial
In-26   1   Seed Farm, Wonotjatur, Kotagede, Bantul, Daera Istimewa Jogjakarta
        2   Middle part of sloping land on footslope colluvium of Mt. Merapi
        3   Regosol
In-27   1   Humo Seed Farm, Semangkak, Kota Klaten, Klaten, Central Java
        2   Middle part of gently sloping land on footslope colluvium of Mt. Merapi
        3   Regosol
In-28   1   Djumapolo, Djumapolo, Karanganjar, Central Java
        2   Middle part of gently sloping land on footslope colluvium of Mt. Lawu
        3   Yellowish brown Latosol
In-29   1   Kalurahan, Sapahan, Sasik Madu, Karanganjar, Central Java
        2   Upper part of gently sloping land on footslope colluvium of Mt. Lawu, resting on consolidated pyroclastic flow
        3   Reddish brown Mediterranean
In-31   1   Kebun Pertjobaan LPPP, Ngale, Paron, Ngawi, East Java
        2   Very gently sloping (S) land on moderately dissected erosional terrace composed of pyroclastic flow
        3   Grumusol
In-32   1   BPMD, Sukodadi, Sukodadi, Lamongan, East Java
        2   Level land on subrecent alluvium (Solo)
        3   Alluvial
In-33   1   BPMD, Brengolo, Kalitidu, Bodjonegoro, East Java
        2   Level land on recent alluvium (Solo)
        3   Alluvial
In-34   1   Krese, Wungu, Madiun, East Java
        2   Middle part of moderately steep land on foot slope colluvium resting on pyroclastic flow of Mt. Wilis
        3   Latosol
In-35   1   Bandjar Sari, Dagangan, Madiun, East Java
        2   Lower part of sloping land on footslope colluvium resting on pyroclastic flow of Mt. Wilis
        3   Brown Mediterranean
In-36   1   Bandjarsari (Patang), Lames, Madiun, East Java
        2   Level (somewhat depressed) land on recent alluvium
        3   Alluvial

In-37 1 Pelem, Paree, Kediri, East Java
       2 Middle part of sloping land on recent footslope colluvium of Mt. Kelud
       3 Regosol

In-38 1 Seed farm, Waung, Baron, Ngandjuk, East Java
       2 Lower part of gently sloping land on recent alluvium
       3 Alluvial

In-39 1 Kebun Pertjobaan, LPPP, Modjosari, Modjosari, Modjokerto, East Java
       2 Lower part of gently sloping land on consolidated pyroclastic flow of Mt. Ardjuno
       3 Ragosol

In-40 1 Seed farm, Bandjarkemantian, Buduran, Sidoardjo, East Java
       2 Level land on recent deltaic alluvium
       3 Alluvial

In-41 1 Maron-kulon, Maron, Probolinggo, East Java
       2 Lower part of gently sloping land on footslope colluvium of Mt. Argopuro
       3 Brown Mediterranean

In-42 1 Labruk Kidul, Lumadjang, Lumadjang, East Java
       2 Level to very gently sloping land on recent alluvial fan deposits resting on pyroclastic flow of Mt. Semeru
       3 Regosol

In-43 1 BPMD, Kalipepe, Jasowilangun, Lumadjang, East Java
       2 Level land on recent alluvium resting on pyroclastic flow of Mt. Lamongan (?)
       3 Alluvial

In-44 1 Seed farm, Srimurin, Ardjasa, Ardjasa, Djember, East Java
       2 Middle part of sloping land on strongly dissected pyroclastic flow on the footslope of Mt. Argopuro
       3 Brown Latosol

In-45 1 Kebun Pertjobaan LPPP, Genteng, Banjuwangi, East Java
       2 Middle part of gently sloping land on consolidated pyroclastic flow of Mt. Raung
       3 Regosol

In-46 1 Seed farm, Sukoredjo, Bangoredjo, Banjuwangi, East Java
       2 Level land on old alluvium
       3 Grumusol

MALAYSIA

West Coast

M-1   1 Kampong Salang, 2 miles north of Kangar, Perlis
      2 Level land on subrecent alluvium
      3 Kangar by McWalter

M-2   1 Kuala Perlis, 0.4 mile south of 6.5 milestone from Kangar, on Kuala Perlis road, Perlis

	2	Depressional level land on recent alluvium (marine)
	3	Kuala Perlis by McWalter
M-3	1	0.5 mile south of Simpang Ampat, 0.25 mile west of the road, Perlis
	2	Level land on recent alluvium (brackish)
	3	Simpang Ampat by McWalter
M-4	1	0.1 mile south of Tanbun Padi Test Station, Perlis
	2	Level land recent alluvium (brackish)
	3	Kampong Telok by McWalter
M-5	1	Langgar, 7.5 miles from Alor Star on Alor Star-Langgar-Tanah Merah road, 0.1 mile west of the road, Kedah
	2	Middle part of gently undulating land on subrecent alluvium
	3	Titi Idris by McWalter
M-6	1	Kuala Kedah, 5.5 miles from Alor Star on Kuala Kedah road, 0.1 mile south of the road, Kedah
	2	Level land on recent alluvium (brackish)
	3	Kuala Kedah by McWalter
M-7	1	Telok Kechai, 5 miles from Alor Star on Kuala Kedah road, 0.1 mile south of the road, Kedah
	2	Level land on recent alluvium (brackish)
	3	Kampong Sedaka by McWalter
M-8	1	Telok Chengai, 0.1 mile south of Telok Chengai Experimental Station, Kedah
	2	Level land on recent alluvium (brackish)
	3	Telok Chengai by McWalter
M-9	1	Guar Chempedah, 18 miles south of Alor Star on the main road, 90 yards east of the road, Kedah
	2	Depressional level land on recent alluvium deposit (brackish)
	3	Yen by McWalter
M-10	1	Sungei Limau, 14 miles south of Alor Star on the main road; 0.5 mile west of the road, Kedah
	2	Level land on recent alluvium (brackish)
	3	Kampong Sedaka series (?) by McWalter
M-11	1	Rantau Panjang, 45.75 miles from Alor Star on Kota Kuala Muda road, Kedah
	2	Level Land on recent alluvium
M-12	1	Bumbong Lima Experimental Station, Province Wellesley
	2	Upper part of gently undulating land on recent levee alluvium
M-13	1	Permatang Benon, 7 miles north northeast of Butterworth, 0.1 mile east of Permatang, Province Wellesley
	2	Middle part of gently undulating land on recent alluvium
M-14	1	Pandang Buloh, between Pandang Manora and Kubang Semang, Province Wellesley
	2	Level land on recent alluvium (partly colluviated?)
M-15	1	Permatang Ara, 6 miles from Butterworth on Kubang Semang road, Province Wellesley
	2	Level land on recent alluvium

M-16 1 Kampong Pertama, 6 miles from Butterworth on Bukit Mertajam road (via Permatang Pauh), 0.25 mile northeast of the road, Province Wellesley
      2 Level land on recent alluvium
M-17 1 Sungei Star Kechil, 0.1 mile east of Sungei Ache Padi Test Station, Province Wellesley
      2 Depressed level land on recent alluvium (brackish)
M-18 1 Tanjong Piandang, 0.1 mile east of Tanjong Piandang Padi Test Station, Krian, Perak
      2 Depressed level land on recent alluvium (marine)
M-19 1 Alor Pongsu, Briah, 0.3 mile southwest of the Padi Test Station, Krian, Perak
      2 Depressed level land on recent alluvium
      3 Briah series
M-20 1 Kampong Mesjid Tingi, 2.5 miles southeast of Bagan Serai, Krian, Perak
      2 Depressed level land on recent alluvium
      3 Braih series
M-21 1 Titi Serong Padi Test Station, Krian, Perak
      2 Level land on recent alluvium (brackish)
M-22 1 Canal 10 road, Lafu Kubong, 6 miles north northeast of Telok Anson, Perak
      2 Depressed level land on recent alluvium
      3 Briah series?
M-23 1 Chenderong Balai, 4 miles west of the junction at 38.5 miles south (near Kampong Degong) of Ipoh, Perak
      2 Upper part of gently undulating land on subrecent alluvium
      3 Manik series?
M-24 1 Chui Chak, 1 mile north of Chui Chak Township, Perak
      2 Level land on flat recent alluvium
M-25 1 Gomback, 5.75 miles north of Kuala Lumpur, Selangor
      2 Level land on subrecent alluvium
M-27 1 Pulau Gadong Padi Test Station, Malaka
      2 Depressed level land on recent alluvium (brackish)
M-28 1 Tedong 12 miles south of Malaka on Muar road, Malaka
      2 Level land on recent alluvium, partly colluviated
M-29 1 Ulu Klawang Padi Test Station, Jelebu, Negri Sembilan
      2 Middle part of gently sloping land on recent alluvium
M-43 1 Block 7, Kian Sit, Sekinchan, Tanjong Karang, Selangor
      2 Level land on recent alluvium (brackish)
      3 Selangor series

East Coast
M-30 1 Kampong Morak, 1.5 miles west of the ferry site, Tumpat, Kelantan
      2 Level land on recent alluvium
M-31 1 Kampong Jambu Ayer, 7 miles south of Tumpat, Kelantan

	2	Level land on recent alluvium
M-32	1	Kampong Manan, 3 miles east of Pasir, Mas, Kelantan
	2	Level land on recent alluvium
M-33	1	Kampong Kadok Dalam, 1 mile east of the 9-mile milestone from Kota Bharu on Kuala Krai road, Kelantan
	2	Level land on recent alluvium
M-34	1	Kampong Padang Salim, Mukim Peringat, 11.5 miles from Peringat on Bachok road, Kelantan
	2	Level land on recent alluvium
M-35	1	Kampong Gaal, Mukim Gaal, 4 miles south of Pasir Puteh, Kelantan
	2	Level land on subrecent alluvium (granite outwash)
M-36	1	Kampong Belukar, Bukit Badak, 2 miles south of Bachok, Kelantan
	2	Upper part of the "bris sand" or subrecent coastal sand ridge
	3	Rudua series
M-38	1	Kampong Kerandang, Kerandang 4 miles east of Jerteh, Trengganu
	2	Level land on subrecent alluvium
	3	Merbau Patah series?
M-39	1	Kampong Sungei Rengas, Glugor, 4 miles southwest of Kuala Trengganu, Trengganu
	2	Level Land on recent alluvium
	3	Tian Bahru series?
M-40	1	Kampong Makemas, Bukit Payong, 9 miles southwest of Kuala Trengganu, Trengganu
	2	Upper part of gently undulating land on subrecent alluvium
	3	Padang Gong Chenak?
M-41	1	Pahang Tua, about 25 miles south of Kuantan along the Pekan road, Kuala Pahang, Pahang
	2	Depressed level land on recent alluvium
	3	Briah?
M-42	1	Kampong Serandu, about 20 miles south of Kuantan along the coast, Kuala Pahang, Pahang
	2	Lower part of the inland side of recent coastal sand ridge
	3	Rusila?

PHILIPPINES

Ph-1	1	Mangahan, Pasigm Rizal (Marikina series)
	2	Level land on subrecent alluvium
Ph-2	1	Biñan, Laguna (Guadalup series)
	2	Middle part of very gently sloping land on subrecent alluvium resting on adobe (hard volcanic flow material)
Ph-3	1	Wakas, Bacaue, Bulacan (Bigaa series)
	2	Level land on subrecent alluvium
Ph-4	1	Santa Rita, Guiguinto, Bulacan (Quingua series)
	2	Level land on recent alluvium (Angat River)
Ph-5	1	Tulao Apalit, Pampanga

	2	Level land on recent alluvium (marine)
	3	San Fernando series
Ph-6	1	Mayondor, Los Baños, Laguna
	2	Upper part of gently undulating land on subrecent alluvium resting on weathered volcanic flow material
	3	Maahas series
Ph-7	1	700 meters south of overpass bridge on highway to Los Baños, 50 meters east of the highway; Balibago, Samta Rosa, Laguna
	2	Middle part of gently sloping land on recent alluvium resting on adobe
	3	Guadalupe series
Ph-8	1	150 meters west of IRRI Farm, 50 meters south of highway; Maahas, Los Baños, Laguna
	2	Lower part of gently sloping land in subrecent alluvium resting on adobe
	3	Maahas series
Ph-9	1	300 meters west of Canlalay junction on highway, Biñan, Laguna
	2	Middle part of very gently sloping land on subrecent alluvium resting on adobe
	3	Carmona series (?)
Ph-10	1	200 meters north of transmitting station, 100 meters west of MacArthur Highway, Bagumbayan, Bacaue, Bulacan
	2	Level land on subrecent alluvium
	3	Bigaa
Ph-11	1	200 meters southeast of Taal Elementary school, Taal, Pulilan., Bulacan
	2	Level land on recent alluvium (Angat River), levee-basin transition
	3	Quingua series
Ph-12	1	200 meters west of Arayat junction on MacArthur Highway; San Augustine, San Simon, Pampanga
	2	Level land on recent alluvium (marine?)
	3	San Fernando series
Ph-13	1	200 meters southeast of bridge about 3.5 km northeast of San Fernando; Lagundi, Mexico, Pampanga
	2	Generally depressed level land on recent alluvium
	3	Quingua (atypical) series
Ph-14	1	100 meters north of road east of Tarlac City; San Manuel, Tarlac, Tarlac
	2	Level to very gently undulating land on recent alluvium resting on older alluvium
	3	Luisita series
Ph-15	1	About 1 km from railroad crossing of road west of Gerona, 50 meters north of the road, Abagon, Gerona, Tarlac
	2	Middle part of gently undulating land on recent alluvium
	3	La Paz series
Ph-16	1	200 meters southwest of elementary school; San Miguel, San

		Manuel Tarlac
	2	Level land on subrecent alluvium
	3	San Manuel series
Ph-17	1	80 meters south of the road from Dagupan to Urdenata; San Jose, Urdenata, Pangasinan
	2	Level to gently undulating land on subrecent alluvium
	3	San Manuel series
Ph-18	1	100 meters east of the road to Baguio; Palacpalac, Pozorrubio, Pangasinan
	2	Middle part of gently sloping land on piedmont plain with recent surface material
	3	Umigan series (?)
Ph-19	1	200 meters east of elementary school; Nancayagan Primero, Urdenada, Pangasinan
	2	Level land on subrecent alluvium
	3	San Manuel series
Ph-20	1	300 meters east of the Rio Chico river-bridge on Guimba—Talavera road; Ilog Baliwag, Sto Rosario, Sto Domingo, Nueva Ecija
	2	Level to very gently undulating land on subrecent alluvium
	3	Quingua series—Maligaya transition?
Ph-21	1	200 meters southeast of elementary school; general A. Luna, Llanera, Nueva Ecija
	2	Middle part of gently undulating land on old alluvium
	3	Prensa series
Ph-22	1	80 meters east of Muños road; Bakal Segundo, Talavera, Nueva Ecija
	2	Middle part of gently sloping land on subrecent alluvium
	3	Maligaya series
Ph-23	1	1 km east of the Chico river on the road from Zaragoza to La Paz; 100 meters north of the road, San Rosario, Zaragoza, Nueva Ecija
	2	Generally depressed level land on subrecent alluvium
	3	Zaragoza series
Ph-24	1	50 meters south of the road; Bertese, Quezon, Nueva Ecija
	2	Level to very gently undulating land on subrecent alluvium
	3	Quingua series
Ph-25	1	70 meters east of the road from Cabanatuan to San Miguel; Sacdalan, San Miguel, Bulacan
	2	Level to very gently undulating land on subrecent alluvium
	3	Bantog series
Ph-26	1	6 km southwest of Cabatuan on San Mateo road, 100 meters north of the road; Diamantina, Cabatuan, Isabela
	2	Level land on recent alluvium (Mangat River)
	3	Sta Rita series
Ph-27	1	4 km west of Cauayan on Cabatuan road, 70 meters north of the road; San Fermin, Cauayan Isabela
	2	Level to very gently undulating land on old alluvium

	3	Cauayan series
Ph-28	1	Southern end of the town of San Manuel (formerly Callang), Isabela
	2	Level to very gently sloping land on subrecent alluvium
	3	Bigaa series
Ph-29	1	120 meters east of Barrio road; Gadu., Solana, Cagayan
	2	Depressional level land on recent alluvium
Ph-30	1	3 km south of Ballesteros, 50 meters west of the road; Mabuttal, Bullisteros, Cagayan
	2	Level land on recent alluvium (marine?)
	3	Zaragoza series (?)
Ph-31	1	150 meters southeast of elementary school; 100 meters east of highway; Toran, Aparri, Cagayan
	2	Level land on recent alluvium
	3	Toran series
Ph-32	1	1 km south of Lallo, 60 meters east of highway; Tukalana, Lallo, Cagayan
	2	Middle part of very gently sloping land on recent alluvium
	3	Quingua series
Ph-33	1	22 km northeast of Iloilo on Highway No. 2, 150 meters west of the road; Nanga, Pototan, Iloilo
	2	Level land on subrecent alluvium
	3	Sta. Rita series
Ph-34	1	1.8 km south of San Miguel, 50 meters west of the road; Poblacion, San Miguel, Iloilo
	2	Middle part of very gently sloping land on subrecent alluvium
	3	Sta. Rita series
Ph-35	1	100 meters southeast of Highway No. 2 junction at Sara; Sara, Iloilo
	2	Very gently undulating land on old alluvium
	3	Sara series
Ph-36	1	1.5 km west of Carmen junction, 50 meters north of the road; Isidoro Fontanilla, Carmen, Davao Norte
	2	Level land on recent alluvium (brackish?)
	3	Cabangan series
Ph-37	1	1.5 km west of Tagum, 100 meters north of the road; Mankilan, Tagum, Davao Norte
	2	Upper part of gently undulating land on recent alluvium
	3	Cabangan series
Ph-38	1	1.5 km west of Kayaga junction, 300 meters south of the road; Malabuaya, Kabacan, Cotabato Norte
	2	Level land on subrecent alluvium
	3	Kabacan series
Ph-39	1	5 km east of Kabacan, 0.3 km south of the road; Katidtuan, Kabacan, Cotabato Norte
	2	Upper part of undulating land on old alluvium

    3   Quilada series

Ph-40 1   2.5 km east of Kabacan, 50 meters south of the road; Katidtuan, Kabacan, Cotabato Norte

    2   Level land on subrecent alluvium

    3   Kabacan series

Ph-41 1   1.5 km south of Bagontapay junction, 80 meters east of the road; Bagontapay, M'lang, Cotabato Sur

    2   Lower part of very gently sloping land on subrecent alluvium

    3   San Manuel series

Ph-42 1   1.6 km southwest of Banga river-bridge, 100 meters west of the road; Liwanay, Banga, Cotabato Sur

    2   Level to gently undulating land on recent alluvium

    3   Banga series

Ph-43 1   300 meters east of Lutayan Municipal Hall; Tamnag, Lutayan, Cotabato Sur

    2   Level land on recent alluvium

    3   Lutayan series

Ph-44 1   400 meters southeast of Barrio 2 schoolhouse; Barrio 2, General Santos City

    2   Upper part of gently sloping land on recent alluvium, partly colluviated

    3   Dadiangas series

Ph-45 1   15 km north of Buluan (or 3 km northeast of Columbio junction, 400 meters east of the road; Dumawato, Columbio, Cotabato Sur)

    2   Level to very gently sloping land on recent alluvium

    3   Timaga series (?)

Ph-46 1   1 km east of Jaro, 100 meters north of the road; Buena Vista, Jaro Leyte

    2   Middle part of gently sloping land on recent alluvial fan deposits

    3   Guinbalaon series

Ph-47 1   6 km east of Alang alang, 50 meters north of the road; Lingayon, Alang alang, Leyte

    2   Level to very gently undulating land on subrecent alluvium

    3   Palo series

Ph-48 1   0.5 km east of Santa Fe, 50 meters north of the road; Pilit, Santa Fe, Leyte

    2   Level to very gently undulating land on subrecent alluvium

    3   Palo series

Ph-49 1   3 km south of Dagami, 50 meters east of the road, Kabuloran, Dagami, Leyte

    2   Middle part of gently sloping land on recent alluvium

    3   Dagami series

Ph-50 1   1 km east of junction to Ocampo, 70 meters north of the road; Sagrada, Pili, Camarines Sur

    2   Middle to lower part of very gently sloping land on old piedmont plain

	3	Pili series
Ph-51	1	150 meters north of Naga City limits, 200 meters west of the road; Haring, Canaman, Camarines Sur
	2	Level land on recent alluvium resting on older alluvium
	3	Pili series
Ph-52	1	0.5 km southeast of Libon on Velarco; 60 meters south of the road; town of Libon, Albay
	2	Level land on recent alluvium
	3	Libon series
Ph-53	1	2.5 km northeast from the junction at Tuburan (east of Ligao), 200 meters north of the road; Batang, Ligao, Albay
	2	Upper part of very gently sloping land on recent alluvium
	3	Ligao series
Ph-54	1	200 meters west of Legaspi Airport gate, 120 meters south of the road; Tagas, Daraga, Albay
	2	Middle part of very gently sloping land on recent alluvium
	3	Legaspi series

SRI LANKA

Sr-1	1	Kaitadi, Jaffna, farmer's land
	2	Depressed level land on recent alluvium
Sr-2	1	Paranthan Paddy Station
	2	Middle part of gently sloping land on subrecent alluvium
Sr-3	1	Muruipu, Mullativu; 28 miles from Mankulam, a farmer's farm
	2	Middle part of undulating land on local alluvium derived from lateritized high terrace sediments
Sr-4	1	Vavunia Agricultural Station
	2	Lower part of gently undulating land on local alluvium derived from gneiss
Sr-5	1	Farmer's field, about 15 km northeast of Mannar
	2	Middle part of gently undulating land on subrecent alluvium, derived partly from limestone
Sr-6	1	Anuradapura Agricultural Station
	2	Middle part of undulating land on local alluvium, derived from gneiss
Sr-7	1	Maha Illupalama Agricultural Station; a field west of the laborers' living quarters
	2	Upper part of gently undulating land on local alluvium, derived from gneiss
Sr-8	1	Murunkan Agricultural Station
	2	Depressional level land on subrecent alluvium
Sr-9	1	Kantalai Paddy Seed Station
	2	Level land on recent alluvium
Sr-10	1	Hingalakgoda Agricultural Station
	2	Middle part of undulating land on local alluvium, derived from gneiss

Sr-11   1   Polonnaruwa Paddy Seed Station
        2   Lower part of gently undulating land on local alluvium, derived from gneiss

Sr-12   1   Pelwehera Farm, Dambula
        2   Upper part of gently undulating land on local alluvium, derived from gneiss

Sr-13   1   Nalanda Paddy Station
        2   Level land on subrecent alluvium, partly derived from limestone

Sr-14   1   Pussellawa Paddy Station
        2   Narrow valley bottom sediments derived from charnockite and gneiss

Sr-15   1   Rahangala, a farmer's field near the government farm
        2   Terraced valley side-slope on gneissic basement

Sr-16   1   Bibile Agricultural Station
        2   Middle part of undulating land on local alluvium, derived from gneiss

Sr-17   1   Karadian Aru Government Farm
        2   Level land on recent sandly alluvium

Sr-18   1   Malvatta Agricultural Farm
        2   Middle part of undulating land on local alluvium, derived from granite

Sr-19   1   Track 1 Inginiagala, a farmer's farm
        2   Lower part of undulating land on local alluvium, derived from gneiss

Sr-20   1   Thotadeniya, Peradeniya Experiment Station
        2   Narrow valley bottom sediments derived from charnockite and gneiss

Sr-21   1   Kundasalle Experiment Station
        2   Narrow valley bottom sediments derived from charnockite

Sr-22   1   Wagulla Farm, Rambukkana, Kegalla
        2   Middle part of undulating land on local alluvium, derived from gneiss

Sr-23   1   Batalagada Central Rice Breeding Station
        2   Upper part of undulating land on local alluvium, derived from gneiss

Sr-25   1   Farmer's field, Minuwangoda
        2   Upper part of gently undulating land on local alluvium, derived from gneiss

Sr-26   1   Farmer's field, Minuwangoda
        2   Lower part of gently undulating land on local alluvium derived from gneiss

Sr-27   1   Bombuwela Rice Research Station
        2   Narrow valley bottom sediments derived from lateritized gneiss

Sr-28   1   Narawela paddy field, Labuduwa Farm, Galle
        2   Level land on recent alluvium

Sr-29   1   Mapalana Government Farm, Kamburupitiya, Matara

	2	Narrow valley bottom sediments derived from gneiss
Sr-30	1	Ambalantota Paddy Station
	2	Upper part of gently undulating land on subrecent alluvium
Sr-31	1	Tissamaharama Paddy Station
	2	Lower part of gently undulating land on recent alluvium
Sr-32	1	Okkampitiya Paddy Station
	2	Middle part of sloping land on local alluvium, derived from gneiss
Sr-33	1	Pottuvill (near Arugam Bay), a farmer's field
	2	Middle part of gently undulating land on subrecent alluvium
Sr-34	1	Karapincha Agricultural Farm
	2	Narrow valley bottom sediments derived from gneiss

THAILAND
Northeast

T-1	1	Ban Tarat, 8 km northeast of Nakhon Ratchasima, Nakhon Ratchasima
	2	Gently undulating land on recent alluvium
	3	Phimai (Pm) series
T-3	1	26 km south of Ban Phai, 300 meters west of Phon–Khon Kaen highway, Khon Kaen
	2	Gently undulating land on semirecent alluvium
	3	Ubon or Roi Et (Sandy Phase) (Ub or Re)
T-4	1	Chum Phae Rice Station, Khon Kaen
	2	Gently sloping land on old alluvium
	3	Phimai series. (Pm)
T-5	1	Ban Don Du, 40 km east of Chum Phae, Khon Kaen
	2	Lower levee part of undulating land on old alluvium
	3	Roi Et (Re) series
T-6	1	Ban Non Mu, 10 km west northwest of Kumphawapi, Udon Thani
	2	Undulating land on old alluvium
	3	Roi Et (Re)
T-12	1	Ban Nong Pleng, Amphoe That Phanom, Nakhon Phanom
	2	Upper level part of undulating land on semirecent alluvium
	3	Roi Et (Loamy Phase) (Re)
T-17	1	Ban Tarom, 3 km southeast of Selaphum, Roi Et
	2	Level land on old alluvium
	3	Roi Et (Loamy Phase) (Re)
T-18	1	18 km east of Roi Et, Roi Et
	2	Level land on old alluvium
	3	Roi Et (Loamy Phase) (Re)
T-22	1	Surin Rice Station, paddy field, Surin
	2	Level land on old alluvium
	3	Roi Et (Loamy Phase ?) (Re)
T-67	1	Phimai Experimental Station, Nakhon Ratchasima
	2	Level land on subrecent alluvium
	3	Phimai (Loamy Phase) (Pm)

T-68   1   Khon Kaen Rice Experimental Station, Khon Kaen
       2   Middle part of gently undulating land on old alluvium
       3   Roi Et (Re)
T-69   1   Ban Kao Noi, 2 km north of Udon, Udon Thani
       2   Level land on old alluvium
       3   Roi Et (Re)
T-72   1   Sakon Nakhon Rice Experimental Station, Sakon Nakhon
       2   Level land on recent alluvium
       3   Udon (Ud)
T-75   1   38 km north of Ubon, Ubon Ratchathani
       2   Middle part of gently undulating land on old alluvium
       3   On (On)
T-76   1   Ban Dong Khaen Noi, Amphoe Kham Khuan Kaeo, Ubon
           Ratchathani
       2   Upper part of gently undulating land on old alluvium
       3   Roi Et (Re)
T-77   1   1 km west of Yasothon, Ubon Ratchathani
       2   Upper part of gently undulating land on old alluvium
       3   Roi Et (Loamy Phase) (Re)
T-78   1   2 km west of Roi Et, Roi Et
       2   Lower part of gently undulating land on old alluvium
       3   Ubon-Roi Et transition (Ub-Re)
T-79   1   Ban Don Khaen, 36.5 km south of Roi Et on Kaset Wisai road,
           Roi Et
       2   Middle part of gently undulating land on old alluvium
       3   Roi Et (Re)
T-80   1   3 km south of Ban Nong Khem, Amphoe Suwannaphum, Roi Et
       2   Level land on recent alluvium, underlain by old alluvium
       3   Tha Tum (Tt)
T-82   1   1 km west of Maha Sarakkham, Maha Sarakham
       2   Upper part of gently undulating land on old alluvium
       3   Roi Et (Re)
T-247 1   Sakon Nakhon (NE 48-10, 0410–1891)
       2   Lower part of undulating land on old alluvium
T-248 1   Nakhon Phanom (NE 48-10, 0478–1917)
       2   Middle part of sloping land on recent levee alluvium
T-251 1   Between Chi and Mun rivers, Si Sa Ket (near Ubon border) (ND
           48-2, 0413–1705)
       2   Level land on recent alluvium overlying a thick pisolith layer
T-252 1   Amphoe Prakhon Chai, Buri Ram (ND 48-5, 0293–1615)
       2   Lower part of sloping land on old alluvium
T-253 1   10 km south of Suwanaphum, Amphoe Suwanaphum, Roi Et (ND
           48-2, 0371–1717)
       2   Level land on recent alluvium, overlying a pisolith layer
T-255 1   1 km south of Nang Rong, Amphoe Nang Rong, Buri Ram (ND
           48-5, 0263–1617)

	2	Lower part of gently sloping land on recent alluvium, underlain by old alluvium
T-9020	1	Pilot Farm Project Huey Si-Thon, Amphoe Muang, Kalasin
	3	Roi Et (Re)
T-9021	1	Pilot Farm Project Huey Si-Thon, Amphoe Muang, Kalasin
	3	Phimai (Pm)
T-9022	1	Amphoe Kang Khro, Chaiyaphoom
	3	US 1 (similar to Ls series)
T-9023	1	North of Ban Du, Amphoe Kanthararam, Si Sa Ket
	3	Roi Et (Re)
T-9024	1	Bhan Ka Tom, Amphoe Muang, Surin
	3	Buriram (Br)
T-9025	1	Ban Dan, Anphoe Sang Kha, Surin
	3	Ubon (Ub)

North

T-28	1	Ban Duang, Amphoe Muang Lamphum, Lamphum
	2	Level land on semirecent alluvium
	3	Lampang (Lp)
T-309	1	Mu Ban Toong Markh Noom, Amphoe Chom Thong, Chiang Mai
	2	Level land on freshly weathered old (Pliocene?) sediment
T-326	1	Mu Ban Chadee, Amphoe Muang, Nan
	2	Middle part of gently undulating land on old alluvium
T-328	1	Mu Ban Sung Men, Amphoe Sung Men, Phrae
	2	Level land on recent alluvium (Yom)

Upper Central Plain

T-152	1	1 km south of Lat Yao, Amphoe Lat Yao, Nakhon Sawan (ND 47-3, 0585–1741)
	2	Middle part of gently sloping land on recent-subrecent fan deposit
T-155	1	Ban Dong Dam, Amphoe Khamu Woralaksaburi, Kamphaeng Phet (NE 47-15, 0580–1773)
	2	Level land on recent alluvium
T-160	1	14 km east of Thung Saliam, Amphoe Thung Saliam, Changwat Sukhothai (NE 47-11, 0579–1914)
	2	Middle part of gently undulating land on old alluvium
T-163	1	1 km east of Si Satchanalai junction, Amphoe Si Satchanalai, Sukhothai (NE 47-11, 0583–1938)
	2	Lower part of sloping land on recent levee alluvium
T-164	1	Ban Huai Rai, Uttaradit (NE 47-11, 0604–1942)
	2	Lower part of gently sloping land on recent alluvium, partially colluviated
T-168	1	Ban Wang Daeng, 2–3 km east of Tron, Amphoe Tron, Uttaradit (NE 47-11, 0621–1933)
	2	Level land on recent alluvium
	3	Ratchaburi (Rb)
T-178	1	8 km south of Sukhothai junction, Sukhothai (NE 47-15, 0586–1873)

	2	Upper part of undulating land on old alluvium
T-183	1	3 km east of Bang Ra Kam, Amphoe Bang Ra Kam, Phitsanulok (NE 47-15, 0624–1855)
	2	Level land on recent alluvium
T-185	1	Ban Wang Chik, 8 km south of Phichit (NE 47-15, 0637–1809)
	2	Level land on recent alluvium
T-194	1	Ban Khao Sai, Amphoe Taphan Hin, Phichit (NE 47-16, 0673–1787)
	2	Upper part of gently sloping land on old alluvium underlain by hard laterite layer
T-196	1	Ban Wang Lum, Amphoe Tapan Hin, Pichit (NE 47-16, 0663–1790)
	2	Lower part of sloping land on recent fan deposit
T-201	1	8 km north of Tha Tako, Amphoe Tha Tako, Nakhon Sawan (ND 47-3, 0657–1738)
	2	Middle part of gently undulating land on old alluvium, Manorom series (Mn)
T-206	1	3.5 km west of Thap Than, Amphoe Thap Than, Uthai Thani, (ND 47-3, 0593–1708)
	2	Recent fan deposit
T-9005	1	Ban Khok Pong, Amphoe Wichianburi, Phetchabun
	3	Wathana (Wa)

Bangkok Plain

T-35	1	3 km east of Ayutthaya on Wang Noi road, Phra Nakhon Si Ayutthaya
	2	Level land on recent alluvium
	3	Bang Khen (Bn)
T-36	1	Ban Pa Ngiu, 3 km northwest of Ang Thong, Ang Thong
	2	Level land on recent alluvium
	3	Ratchaburi (Rb)
T-38	1	Ban Bung, 6 km west of Tha Rua, Phra Nakhon Si Ayutthaya
	2	Gently sloping land on recent alluvium
	3	Bang Khen (Bn)
T-43	1	Ban Tha Ngam, Amphoe In Buri, Sing Buri
	2	Level land on semirecent alluvium
	3	Ratchaburi (Rb)
T-46	1	Ban Thuk, Amphoe Sanphaya, Chai Nat
	2	Level land on semirecent alluvium
	3	Ratchaburi (Rb)
T-60	1	Ban Samrong, Amphoe Phra Khanong, Phra Nakhon
	2	Level land on recent alluvium (marine)
	3	Bangkok-Bang Khen transition (Bk-Bn)
T-63	1	Ban Lat Bua Luang, Amphoe Lat Bua Luang, Phra Nakhon Si Ayutthaya
	2	Level land on recent alluvium (brackish)
	3	Ongkharak (Ok)

T-83 1 Bang Khen Experimental Station, Changwat Phra Nakhon
      2 Level land on recent alluvium (marine?)
      3 Bang Khen (Bn)

T-84 1 Rangsit Experimental Station, Pathum Thani
      2 Level land on recent alluvium (brackish)
      3 Ongkharak (Ok)

T-88 1 Tambon Ban O, Amphoe Ban Na, 15 km toward Ongkharak on Nakhon Nayok Ongkharak road, Nakhon Nayok
      2 Level land on recent alluvium (brackish)
      3 Rangsit (Rs)

T-91 1 Ban Kachap, Amphoe Ban Pong, 12 km west of Nakhon Pathom, Ratchaburi
      2 Level land on recent alluvium
      3 Ratchaburi (Rb)

T-95 1 Ban Song Khlong, 63 km from Bangkok on Chon Buri road, Chachoengsao
      2 Level land on recent alluvium (marine)
      3 Samut Prakan (Sm)

T-111 1 13 km east of Min Buri, Amphoe Min Buri, Phra Nakhon (ND 47-12, 0701–1526)
      2 Level land on recent alluvium (brackish)

T-115 1 North of Amphoe Phanom Sarakham, Chachoengsao (ND 47-12, 0754–1522)
      2 Upper part of gently sloping land on recent local alluvium, derived from laterite-bearing hills

T-124 1 15 km southeast of Prachin Buri on the road to Si Maha Phot, Amphoe Si Maha Phot, Prachin Buri (ND 47-12, 0766–1547)
      2 Level land on recent alluvium

T-128 1 10 km southwest of Prachin Buri, Prachin Buri (ND 47-12, 0748–1549)
      2 Level land on recent alluvium (brackish)

T-132 1 2 km east of Ban Ko Raet, Amphoe Bang Len, Nakhon Pathom (ND 47-11, 0627–1546)
      2 Level land on recent alluvium, underlain by old alluvium

T-134 1 Ban Suan Taeng, along the road from U-Thong to Suphan Buri, Suphan Buri (ND 47-7, 0607–1596)
      2 Level land on recent alluvium with gypsum (brackish)

T-139 1 5 km east of Sawaeng Ha, Amphoe Sawaeng Ha, Ang Thong (ND 47-7, 0647–1631)
      2 Upper part of gently undulating land on old alluvium

T-147 1 5 km west of Lop Buri, Lop Buri (ND 47-8, 0670–1638)
      2 Depressed levee land on recent alluvium

T-212 1 13 km south of Photharam, Ratchaburi (ND 47-11, 0593–1506)
      2 Level land on recent alluvium with gypsum (brackish)

T-213 1 5 km northwest of Phetchaburi, Phetchaburi (ND 47-11, 0598–1456)

        2   Level land on recent alluvium
T-216   1   1.5 km south of Bang Bua Thong, Amphoe Bang Bua Thong,
            Nonthaburi (ND 47-11, 0645–1538)
        2   Level land on recent alluvium with gypsum (brackish)
T-244   1   Ban Yi Hon, Amphoe Bang Pla Ma, Suphan Buri (ND 47-7,
            0628–1591)
        2   Depressed level land on recent alluvium (brackish)

Southern Peninsula
T-224   1   6 km southeast of Sungai Kolok junction, Amphoe Sungai Kolok
            (?) Narathiwat
        2   Lower part of sloping land on subrecent alluvium
        3   Bangnara (Ba)
T-226   1   9.5 km southwest of Pattani Bridge, Pattani
        2   Lower part of sloping land on recent alluvium (brackish)
T-232   1   4 km west of Ban Lam Pam, Amphoe Muang, Phatthalung (ND
            47-3, 0624–0843)
        2   Level land on recent alluvium
T-234   1   4 km west of railway crossing at Ban Nong Taen, Nakhon Si
            Thammarat (NC 45-17, 0623–0921)
        2   Lower part of gently sloping land on subrecent alluvium
T-237   1   5 km east of Nakhon Si Thammarat-Pak Phanang junction, Nakhon
            Si Thammarat (NC 47-15, 0612–0920)
        2   Level land on subrecent alluvium
T-240   1   Ban Khlong Noi, Amphoe Muang, Surat Thani (NC 47-11,
            0535–1015)
        2   Level land on recent alluvium

VIETNAM
V-1   1   4 km south of Ke Sach, 50 meters west of road, Phu Tam, Thuan
            Hoa, Ba Xuyen
        2   Depressed part of tidal flat (1.2.1)*
V-2   1   500 meters west of Dinh Thanh Village Office, Dinh Thanh, Quan
            Long, An Xuyen
        2   Depression on high tidal flat (1.1.3)
V-3   1   5 km west of Tac Soy, An Trach, Gia Rai, Bac Lieu
        2   Low tidal flat (1.2.3), shallow flood
V-4   1   500 meters east of Long Thanh bridge, Tra Kha A hamlet, Long
            Thanh, Vinh Loi, Bac Lieu
        2   High tidal flat (1.1.2), shallow flood
V-5   1   300 meters east of Vinh An, Vinh Trach, Vinh Loi, Bac Lieu
        2   Low tidal flat (1.2.3), shallow flood
V-6   1   300 meters northeast of Thanh Tri Bridge, Thanh Tri, Thanh Tri,
            Ba Xuyen
        2   High tidal flat (1.1.2), medium flood

*Physiographic unit number according to the Netherlands Delta Development
Team (1974).

V-7	1	300 meters south of the road, 3.5 km southwest of road junction at Da Tam (edge of a sand ridge), Da Tam, My Xuyen, Ba Xuyen
	2	High tidal flat (1.1.2), shallow flood
V-9	1	300 meters west of An Ninh Village Office, 100 meters north of the road, An Ninh (or Ba Thao), Thuan Hoa, Ba Xuyen
	2	High tidal flat (1.1.1), medium flood
V-10	1	About 1 km southeast of Tiep Nhut Elementary School, Vien An Lich Hoi Thuong, Ba Xuyen
	2	High tidal flat (1.1.1), shallow flood
V-12	1	Phuoc Hoa, Phu Tam, Thuan Hoa, Ba Xuyen, demonstration farm south of the village
	2	High tidal flat (1.1.1), shallow flood
V-13	1	Soc Trang, Phung Hiep, Phung Hiep, Phong Dinh
	2	Backswamp of estuarine flood plain (2.2.1), intermediate flooding
V-14	1	Xeo Vong B hamlet, Phung Hiep, Phung Hiep, Phong Dinh
	2	Backswamp of estuarine flood plain (2.2.1), intermediate flood
V-15	1	200 meters east of road, about 200 meters west of Tan Thanh bridge, Tan Thanh, Chan Than, Phong Dinh
	2	Backswamp of estuarine flood plain (2.2.1), intermediate flood
V-16	1	Binh Hoa, Binh An, Phong Phu, Pong Dinh
	2	Backswamp of estuarine flood plain (2.2.1), intermediate flood shallow
V-17	1	Thoi Hoa, Thoi Thanh, Phong Pha, Phong Dinh
	2	Broad depression, acid soil (4.1.4) deep flood
V-18	1	3 km west of Bon Tong Canal, My Thanh, Dinh My Hue Duc, An Giang
	2	Broad depression, acid soil (4.1.4), deep flood
V-19	1	Than Hue, My Thoi, Chau Thanh, An Giang
	2	Backswamp of estuarine floodplain (2.2.3), deep flood
V-20	1	About 4 km from JCT; Vinh Thanh, Vinh Trinh, Thot Not, An Giang
	2	Backswamp of estuarine floodplain, deep flood (2.2.3)
V-21	1	3 km west of Kien Tan along the Kien Tan Canal: Tan Hoa, Tan Hiep, Kien Tan, Kien Giang, the edge of the demonstration farm
	2	Broad depression, acid soil (4.1.4)
V-22	1	South of demonstration farm: Hoa An, An Hoa, Rach Gia, Kien Giang
	2	High tidal flat (1.1.2)
V-23	1	Binh Phu, Binh Hoa, Chau Thanh, An Giang
	2	Levee of estuarine floodplain (2.1.2) intermediate flood
V-24	1	About 1 km east of Tinh Bien, Phu Hun, Xung To, Tinh Bien, Chau Doc
	2	Piedmont plain (7)
V-25	1	About 500 meters east of canal bridge (nearest to the hill), Son Hoa, Thoi Son, Tinh Bien, Chau Doc
	2	Broad depression, very acid soil (4.2.3), deep flood
V-26	1	About 1 km west of 230 meters hill, Vinh Khanh, Vinh Te, Chau

               Phu, Chau Doc

	2	Broad depression, acid soil (4.1.4), deep flood
V-27	1	Chau Thoi 2 hamlet, Chau Phu, Chau Phu, Chau Doc
	2	Backswamp of estuarine floodplain (2.2.3) deep flood
V-28	1	2.5 km from ferry, Vinh Tuong, Chau Phong, Chau Phu, Chau Doc
	2	Levee to backswamp transition of riverine floodplain (3.1–3.2)
V-29	1	4 km west of Tan Chau; Long Qoui A hamlet, Long Phen, Tan Chau, Chau Doc
	2	Levee to backswamp transition of river floodplain (3.1–3.2)
V-30	1	Phu Xuong, Phu An, Tan Chau, Chau Doc
	2	Backswamp of estuarine floodplain (2.2.3)
V-31	1	4 km from Vinh Thanh; Hung Thanh Tay, Long Hung, Lap Vo, Sadec
	2	Backswamp of estuarine floodplain (2.2.3) deep flood
V-32	1	About 6 km from Sadec; Tan Binh, Hoa Thanh, Duc Thanh, Sadec
	2	Backswamp of estuarine floodplain (2.2.1)
V-33	1	About 1 km north of the road, Phong Phu, Phong My, Cao Lanh, Kien Phong
	2	Levee to backswamp transition (2.1.2–2.2.3) intermediate flood
V-34	1	About 3 km from Cao Lanh: Tinh Chau, Tinh Thoi, Cao Lanh, Kien Phong
	2	Levee of estuarine floodplain (2.1.2), intermediate flood
V-35	1	Tan Phu, An Nhom, Duc Ton, Sadec
	2	Levee of estuarine floodplain (2.1.1), intermediate flood
V-36	1	About 4 km north of T junction, Phu Hoa, Phong Thanh, Cao Long, Vinh Binh
	2	Low tidal flat (1.2.1), intermediate flood
V-37	1	165 km milestone, Nhon Ngai, Hien Phung, Vung Liem, Vinh Long
	2	Backswamp of estuarine floodplain (2.2.1) intermediate flood
V-38	1	155 km milestone; Ngai, Tan Hoi, Minh Duc, Vinh Long
	2	Broad depression acid soil (4.1.1), intermediate flooding
V-39	1	146.5 km milestone: An Loung, An Duc, Chau Thanh, Vinh Long
	2	Backswamp of estuarine floodplain (2.2.1)
V-40	1	149 km milestone: Tan Phu, Song Phu, Binh Minh, Vinh Long
	2	Broad depression acid soil (4.1.2)
V-41	1	300 meters west of Cai Von Bridge, My Thoi, My Thuan, Binh Minh, Vinh Long
	2	Backswamp of estuarine floodplain (2.2.1)
V-42	1	117 km milestone; Ap Cho. An Thai Dong, Goa Duc, Dinh Thong
	2	Backswamp of estuarine floodplain (2.2.1), shallow flood
V-43	1	91 km milestone (1 km east of Cai Lay); Nhi My, Nhi My, Cai Lay, Dinh Tuong
	2	Backswamp of estuarine floodplain (2.2.1), shallow flood
V-44	1	68 km milestone; Long Khuong, Long An, Chau Thanh, Dinh Tuong
	2	High tidal flat (1.1.1), shallow flood

V-45 1 31 km from Trac Giang; An Dinh, An Ngai Trung, Ba Tri, Kien Hoa
      2 Low tidal flat (1.2.1), intermediate flood
V-46 1 9.5 km from Trac Giang, Phong Dien, My Thanh, Giang Trom, Kien Hoa
      2 High tidal flat (1.1.1), shallow flood
V-47 1 80 km milestone from Saigon; Ap 4, Hien Dinh, Truc Giang, Kien Hoa
      2 Low tidal flat (1.2.2), mainly acid, shallow-intermediate flood
V-48 1 At the sharp bend of road, Ap Tu, Binh Thanh Dong, Hoa Tan, Go Cong
      2 Low tidal flat (1.2.1), shallow-intermediate flood
V-49 1 33 km east of My Tho Long Hung, Long Thuan, Hon Lac, Go Cong
      2 High tidal flat (1.1.1), shallow flood
V-50 1 10 km east of My Tho Long Thanh, Long Binh Dien, Cho Gao, Dinh Tuong
      2 High tidal flat (1.1.1), shallow flood
V-51 1 41 km milestone; Binh Cau, Binh Phong Thanh, Thu Thua, Long An
      2 High tidal flat (1.1.1), shallow flood

Appendix 2
Methods of Analysis

1. pH

A 10 g sample of air-dried fine soil (< 2 mm) was suspended in 50 ml of deionized water ($< 1\mu$mho/cm). After standing for about an hour the mixture was stirred and the pH of the suspension was determined with a glass electrode pH meter.

2. Total Carbon

A modified wet combustion method after Tyurin was adopted. Phenylanthranilic acid was used as the indicator in the titrimetry, instead of diphenylamine.

3. Total Nitrogen

Kjeldahl digestion and steam distillation were adopted in most cases. Occasionally a microdiffusion method was used instead of steam distillation.

4. Available Nitrogen or Ammoniacal Nitrogen Produced in Anaerobic Incubation

A 10 g sample of air-dried fine soil was weighed into a glass cylinder ($\phi = 22$ mm and height $= 140$ mm). 25 ml of water was added to submerge the soil. After thorough stirring to remove trapped air, the soil was incubated at 40°C for two weeks or occasionally for four or six weeks. The supernatant was then filtered and 25 ml of 1N KCl solution was added to the soil. After thorough stirring, the mixture was left to stand for about 30 minutes. The supernatant was filtered and another 25 ml of 1N KCl solution was added to the soil. After stirring, the soil was transferred to filter paper and washed ten times with 5 ml of 1N KCl solution. Filtrates and the washings were made to an appropriate volume, from which an aliquot was taken for NH_3-N determination by either steam distillation or the microdiffusion method.

5. Total Phosphorus

A 100 mg finely ground air-dried soil sample was placed in a Pt crucible, and 3 ml 6N H_2SO_4 solution and 3 ml concentrated HF solution were added. The mixture was heated to dryness and the organic matter was decomposed by ignition. The HF-H_2SO_4 treatment was repeated. Five ml of 6N H_2SO_4 solution was added and the mixture was heated to dissolve the residue. From this solution phosphorus was determined by colorimetry with molybdenum yellow.

6. Bray-Kurtz Soluble Phosphorus

The Bray-Kurtz No. 2 method was used with a few modifications.

7. *0.2N HCl Soluble Phosphorus*

Fifty ml of 0.2N HCl solution were added to 2.5 g of air-dried fine soil in a 100 ml Erlenmayer flask. The soil-solution mixture was kept for 5 hours in a 40°C incubator and handshaken every hour. The solution was then filtered and phosphorus was determined by colorimetry with molybdenum blue in 1:10 HCl medium.

8. *CEC*

Forty ml of 1N $CaCl_2$ solution buffered with triethanolamine to pH 8.2, was added to a 4 g sample of air-dried fine soil. After standing for 30 minutes with occasional stirring, the suspension was filtered and the residual soil washed until the filtrate was free of chloride ion. The soil was then washed twice with 40 ml 1N NaCl, pH = 7. The Ca ion content in the washing was determined by titrimetry with EDTA, and the result was expressed in me/100 g.

9. *Exchangeable Cations*

For Ca^{2+} and Mg^{2+} a sample of air-dried fine soil was extracted twice with an 1N NaCl solution in a soil-solution ratio of 1:10. The cations were determined by EDTA titrimetry. For Na^+ and K^+, a sample of air-dried fine soil was extracted twice with 1N CH_3COONH_4 solution in a soil-solution ratio of 1:10. The cations were determined by flame photometry.

10. *Available Silica*

Using 1N CH_3COOH, pH 4.0 (adjusted with 1N CH_3COONa solution), a sample of air-dried fine soil was extracted in a soil-solution ratio of 1:10 at 40°C for five hours with occasional shaking. The silica extracted was determined by colorimetry with molybdenum blue in a 1:10 HCl medium.

11. *Free Oxides*

One hundred ml 0.2N oxalic acid solution was added to a 1 g sample of air-dried fine soil in a 100 ml Erlenmeyer flask. The flask was exposed to sunlight for about five hours and shaken by hand every hour. The time required for photochemical reduction fluctuated depending on light intensity, air-temperature, etc. Therefore, a standard sample was always included in a lot to control the accuracy of the method (coefficient of variance 9 percent for reproducibility, 3 percent for repeatability). After photochemical reduction the sample was left overnight, then washed into a 250 ml volumetric flask containing 35 ml of a 1:2 HCl solution. The solution was made up to volume with deionized water and thoroughly shaken. After standing for more than 1 hour, aliquots of the clear supernatant were taken for determination of the following elements:

Fe	Colorimetry with *o*-phenanthroline
Mn	Atomic absorption spectrometry
Al	Colorimetry with mercaptoacetic acid-aluminon after decomposing the oxalic acid.

12. *Phosphorus Absorption Coefficient*

A 50 g sample of air-dried soil was shaken with 100 ml of a 25 percent $(NH_4)_2HPO_4$ solution with pH adjusted to 7.0. After twenty-four hours reaction, an aliquot of the filtrate was taken for colorimetric determination of phosphorus. The amount of phosphorus absorbed by the soil was calculated from the difference between the original amount and the remaining amount of phosphorus in solution. The phosphorus absorption coefficient was expressed as mg of P_2O_5 absorbed by 100 g of oven-dried soil.